Control Theoretic Splines

Optimal Control, Statistics, and Path Planning

PRINCETON SERIES IN APPLIED MATHEMATICS

Series Editors: Ingrid Daubechies (Princeton University); Weinn E (Princeton University); Jan Karel Lenstra (Eindhoven University); Endre Sli (University of Oxford)

The Princeton Series in Applied Mathematics publishes high-quality advanced texts and monographs in all areas of applied mathematics. Books include those of a theoretical and general nature as well as those dealing with the mathematics of specific applications areas and real-world situations.

Titles in the series

Chaotic Transitions in Deterministic and Stochastic Dynamical Systems Applications of Melnikov Processes in Engineering, Physics, and Neuroscience, by Emil Simiu

Self-Regularity A New Paradigm for Primal-Dual Interior Point Algorithms, by Jiming Peng, Cornelis Roos, and Tamas Terlaky

Selfsimilar Processes, by Paul Embrechts and Makoto Maejima

Analytic Theory of Global Bifurcation: An Introduction, by Boris Buffoni and John Toland

Entropy, edited by Andreas Greven, Gerhard Keller, and Gerald Warnecke

Auxiliary Signal Design for Failure Detection, by Stephen L. Campbell and Ramine Nikoukhah

Max Plus at Work Modeling and Analysis of Synchronized Systems: A Course on Max-Plus Algebra and Its Applications, by Bernd Heidergott, Geert Jan Olsder, and Jacob van der Woude

Optimization: Insights and Applications, by Jan Brinkhuis and Vladimir Tikhomirov

Thermodynamics: A Dynamical Systems Approach, by Wassim M. Haddad, Vijay-Sekhar Chellaboina, and Sergey G. Nersesov

Impulsive and Hybrid Dynamical Systems Stability, Dissipativity, and Control, by Wassim M. Haddad, VijaySekhar Chellaboina, and Sergey G. Nersesov

Genomic Signal Processing, by Ilya Shmulevich and Edward Dougherty

Positive Definite Matrices, by Rajendra Bhatia

The Traveling Salesman Problem: A Computational Study, by David L. Applegate, Robert E. Bixby, Vasek Chvatal, and William J. Cook

Wave Scattering by Time-Dependent Perturbations: An Introduction, by G. F. Roach

Algebraic Curves over a Finite Field, by J.W.P. Hirschfeld, G. Korchmros, and F. Torres

Control Theoretic Splines

Optimal Control, Statistics, and Path Planning

Magnus Egerstedt and Clyde Martin

PRINCETON UNIVERSITY PRESS

PRINCETON AND OXFORD

Published by Princeton University Press,

41 William Street, Princeton, New Jersey 08540

In the United Kingdom: Princeton University Press,

6 Oxford Street, Woodstock, Oxfordshire OX20 1TW

Library of Congress Cataloging-in-Publication Data

Egerstedt, Magnus.
    Control theoretic splines : optimal control, statistics, and path planning / Magnus
Egerstedt and Clyde Martin.
    p. cm. – (Princeton series in applied mathematics)
Includes bibliographical references and index.
ISBN 978-0-691-13296-9 (hardcover : alk. paper) 1. Interpolation. 2. Smoothing
(Numerical analysis) 3. Smoothing (Statistics) 4. Curve fitting. 5. Splines. 6.
Spline theory. I. Martin, Clyde. II. Title.
QA297.6.E44 2010
511'.42–dc22

                          2009034177

The publisher would like to acknowledge the authors of this volume
for providing the camera-ready copy from which this book was printed.

British Library Cataloging-in-Publication Data is available

Printed on acid-free paper. ∞

press.princeton.edu

Printed in the United States of America

10 9 8 7 6 5 4 3 2 1

To our wives Danielle and Joyce
and
to our friend and mentor Roger Brockett.

Contents

PREFACE

In 1998, Clyde Martin visited the Royal Institute of Technology in Stockholm and taught a course on advanced topics in systems theory. Among the students were Magnus Egerstedt, and what started as a homework assignment quickly led to the discovery that it was possible to generalize the smoothing splines concepts, as defined by Grace Wahba in the area of statistics, using standard control theoretic ideas. The key enabling (yet rather obvious) observation was that a rich class of smoothing curves can be traced by the output of a linear control system, driven by an appropriately selected input. However, that the corresponding class of curves captured almost all of the traditional splines, as well as leading to useful new areas of investigation, including monotone splines, splines with continuous data, and splines on manifolds, was more of a surprise. During the last ten years, a rather large body of work has been developed, connecting splining concepts to those found in the systems theory literature. This book is the outcome of that study.

Everyone who reads this book will realize that the basic material owes a great deal to the "red book" of David Luenberger. His concept of optimization using vector space methods is one of those ideas that has had a major influence in engineering, economics, mathematics, and every other area that is concerned with basic optimization. Both authors have taught from and have been taught from his basic book. We have strived to bring to this book some of the readability properties that David has mastered, as well as to connect with the tools and techniques developed by him.

Although the material in this book covers a lot of ground, a word of caution is in order. We have made no attempt to survey the huge field of splines, even of smoothing splines. The purely statistical approach to smoothing splines differs from our approach in the end application, but there is a huge overlap in basic concepts. Anyone who is interested in the statistical approach to smoothing splines is urged to read the seminal monograph of Grace Wahba and the excellent monograph of Randy Eubanks on this topic.

The material in this book relies heavily on a very fruitful collaboration with Professor Yishao Zhou, and it is fair to say that the book would not have been the same without her. In particular the chapter on smoothing splines as integral filters owes much to her and to Professor W. P. (Daya) Dayawansa. Daya has contributed to the content of the book, but, more important, he has been a pioneer in the interface between mathematics and engineering. Both authors are indepted to him for his influence on their scientific philosophy.

The authors have also been heavily involved with Professor Hiroyuki Kano in the development of applications of B-splines. We decided not to include that material in this book because of our focus on curves generated by linear control systems with a particular optimization format. However, there certainly is some overlap in applications areas that the interested reader is urged to explore through the extensive publications of Professor Kano.

When developing smoothing splines, statistics cannot be ignored. The second author was patiently reminded to use statistics correctly by Shan Sun. Professor Sun was a coauthor on the first two major papers in the development of control theoretic smoothing splines. When statistics are used correctly, she deserves the credit, while the authors assume the responsibility for all statistical bobbles.

The very first paper in this series was a collaborative effort with Professor Zhimin Zhang. That paper mimicked the classical spline construction, and the approach used was quite different from the approach used in this book. Nonetheless, it served as a starting point for our study of the connection between splines and linear systems theory.

There have been many graduate students involved with the development and application of control theoretic splines at Texas Tech University, Royal Institute of Technology in Stockholm, Georgia Institute of Technology, and Stockholm University. Many of their names can be found on papers in the bibliography. We are so very grateful for the work that they have done on this long term project. We thank the past, present, and future students for their diligence.

Modern research cannot be done without financial support. We have been fortunate to have been supported by many agencies: NSF, NASA, NSA, AFOSR, ARO, EPA, NIH, and DARPA. We gratefully acknowledge the support these agencies have provided over the years.

On a personal note, we want to thank our families and friends for supporting our work. In particular, Danielle Hanson has been a constant source of joy, energy, and inspiration to Magnus. Not only has she kept Magnus's mind (somewhat) straight in terms of providing a big picture, but she has also been involved in many technical discussions at the dinner table. Thank you!

Joyce Martin has stood beside Clyde for 45 years and has never flinched. She has understood when mathematics was first on his mind, and she has patiently stood by as he traveled even when there were four small children at home. She has always been ready to be the occasional pro bono editor of his papers and books. Not only does she deserve the credit for this book but for all of the work that Clyde has done!

Atlanta and Lubbock – February 2009

Chapter One

INTRODUCTION

Splines are ubiquitous in science and engineering. Sometimes they play a leading role as generators of paths or curves, but often they are hidden inside, for example, software packages for solving dynamic equations, in graphics, and in numerous other applications.

The standard, classic spline is an interpolating curve. In contrast to this, smoothing splines are only required to pass "close" to the data points. Such smoothing splines are well know by name in statistics, but not so well known outside of this area. The goal of this book is to show that smoothing splines arise as a natural part of control theory, and that, by using control theoretic concepts, we can construct and interpret smoothing splines in an efficient, algorithmic manner.

Throughout the book, this connection between control theory and smoothing splines will be made explicit, and we will find numerous applications for smoothing splines in path planning for mobile robots, in numerical analysis, graphics, and other basic applications. This introductory chapter presents a brief background to interpolating and smoothing splines, as well as sets up their connection to linear systems theory.

1.1 FROM INTERPOLATION TO SMOOTHING

The basic problem that the classical spline was constructed to solve was as follows: Given a finite set of data points, find a smooth curve that interpolates through these points. Of course, there are infinitely many such curves, and the real task is to devise an algorithm that selects a unique (hopefully exhibiting certain desirable properties) curve. In fact, classical splines solve this problem by requiring that the curve be piecewise polynomial, that is, that it be polynomial between the data points, and that the pieces be connected as smoothly as possible. Often additional conditions must be applied as well at the endpoints to ensure uniqueness.

This idea of producing interpolating polynomials, stitched together at the data points, works wonderfully if the data are exact, or nearly so. Unfortunately, data often have significant error associated with them, and classical splines tend to accent these errors. *Smoothing splines* were developed to remedy this very problem, that is, to handle cases when there is error associated with the data points. Naturally enough, these smoothing splines were developed in statistics, where noise is a fact of life, and where error is assumed in almost all data. As such, the restriction of exact interpolation was dropped, while the restriction remained that the curves should be piecewise polynomial and as smooth as possible.

Statistics aside, this notion of producing smoothing rather than interpolating curves is rather natural as well in engineering in general, and control theory in particular. In fact, various notions of controllability have always played fundamental roles in engineering through the canonical problem of moving an object at a known position with known dynamics to a new position. For example, in air traffic control, ground control typically will dictate to the pilot of an airplane where it should be at a fixed set of times, and what its corresponding directions should be, for example, the command could be to be at 10,000 feet in 2 minutes with a given heading. The pilot will in fact receive a string of such commands as the plane approaches an airport. Typically, some deviations from the exact locations are allowed, and the size of the deviation depends on many factors. For example, passenger comfort requires that accelerations are minimized, and that transitions are smooth. As a consequence, exact interpolation is not desirable in this case. In fact, the pilot is constructing a type of smoothing spline.

Based on this rather informal observation, it seems natural to give a more explicit description of the general controllability problem in the context of smoothing splines. It was from this rather straightforward idea that the concept under investigation in this book arose, that is, the concept of control theoretic splines.

1.2 BACKGROUND

The problem of approximation is almost as old as modern mathematics. In fact, polynomial interpolation dates back to the mid 1700s, with the work of Edward Waring (Lagrange interpolation). The ideas of polynomial approximation were central during the 1800s, with the development of various families of orthogonal polynomials, and what later became known as the related Hilbert space theories. The polynomial interpolation problems were of such importance that a significant part of modern mathematics can trace its history back to these developments in one form or another. But, if polynomial interpolation is such a well-studied and powerful tool, then why were polynomial splines invented?[1]

---

[1] By splines, we here mean piecewise polynomial curves that are stitched together at given nodal points in order to ensure certain regularity properties.

1.2.1 Polynomial Interpolating Splines

Traditional (pre-spline) polynomial interpolation has at least two very serious drawbacks, which limit its use in many applications. The first is that a polynomial of degree $n + 1$ may have as many as n local extrema. This causes the curve to be very complex. If, for example, we have $n + 2$ data points that are connected by curves that are approximately linear, then the interpolating polynomial will have degree $n + 1$, and hence will not at all be approximately (piecewise) linear. As such, while we may have a locally good fit, we cannot have a good fit over an arbitrarily large interval.

The second major drawback is an algorithmic problem. To find a polynomial that interpolates a given set of data is equivalent to inverting a van der Monde matrix. The condition number of a van der Monde matrix can grow as 2^{2^N}, with N being the size of the matrix, making the inversion rather intractable in that numerically, the problem of polynomial interpolation may become highly unstable (see e.g., [42]). So, as beautiful as the theory of polynomial interpolation is, it is not particularly useful for large problems.

To remedy this, during the early 1940s, splines as we know them were invented by Isaac Schoenberg at the U.S. Army Ballistic Research Laboratory in Aberdeen, Maryland (the Aberdeen Proving Ground). The splines' early uses are somewhat shrouded in mystery, as this was highly classified research, and it was not until after the Second World War that Schoenberg publicly described his invention.

Schoenberg formulated the spline problem in the following manner. Let $D = \{(t_i, \alpha_i) : i = 1, \ldots, n\}$ be a set of time-stamped data points (with t_i the time stamp and α_i the data point), and let F be the set of twice continuously differentiable functions that interpolate the data, that is, $F = \{f \in C^2[0, T] \mid f(t_i) = \alpha_i\}$. Now, the spline problem is given in terms of the following optimization problem:

$$\min_{f \in F} \max_{t \in [0,T]} |f''(t)|,$$

where f'' denotes second derivative. What this problem entails is to find the interpolating function f that has the smallest maximal second derivative on the interval in question.

While this formulation is very elegant, it is (at least at first glance) not an easy problem to solve. In fact, this formulation constitutes an optimization problem over a notoriously difficult Banach space–the space of continuous functions on a compact set. Luckily, the solution is the classical *cubic spline*.

The observation that the cubic spline (piecewise polynomial curves of degree three) solved Schoenberg's problem led to the development of a host of good numerical algorithms for the construction of the optimal solution, based explicitly on the cubic nature of the solution polynomials. As a consequence, the focus shifted to an in-depth study of such piecewise (cubic) polynomials, while the original optimization problem was largely ignored for nearly three decades. A comprehensive overview of classical splines can be found in Carl de Boor's *A Practical Guide to Splines* [24].

1.2.2 Polynomial Smoothing Splines

It was not until the early 1970s that Grace Wahba (who appropriately enough happens to be the I. J. Schoenberg Professor of Statistics at the University of Wisconsin-Madison) began to study the use of splines with noisy data, that the underlying optimization problem was revisited. In fact, one of Wahba's most important contributions to the subject was to replace the Banach space problem with the much simpler Hilbert space problem

$$\min_{f \in L_2[0,T]} \int_0^T f''(t)^2 dt + \lambda \sum_{i=1}^n (f(t_i) - \alpha_i)^2. \tag{1.1}$$

Here L_2 denotes the Hilbert space of square integrable functions, and $\lambda > 0$ is a weight that determines the tradeoff between the smoothness of the solution and the closeness between curve and data points. An example of interpolating and smoothing curves, as formulated by Schoenberg and Wahba, is given in Figures 1.1 and 1.2.

1.3 THE INTRODUCTION OF CONTROL THEORY

It should be noted already at this point that the formulation in (1.1) requires a certain leap of faith, since most L_2 functions are not differentiable, that is, the second derivative, f'', may not be well defined. However, this small inconvenience can be easily remedied by the use of a little control theory.
   Let

$$f''(t) = u(t),$$

and let

$$y(t) = \int_0^t (t - s)u(s)ds. \tag{1.2}$$

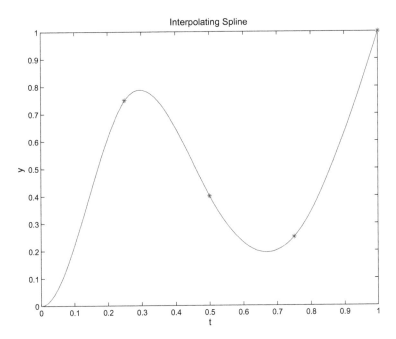

Figure 1.1 Interpolating cubic splines.

Then Wahba's optimization problem can be reformulated as

$$\min_{u \in L_2[0,T]} \int_0^T u(t)^2 dt + \lambda \sum_{i=1}^n (y(t_i) - \alpha_i)^2.$$

Based on this formulation, we are only one small step away from the full-fledged control theoretic formulation that will be pursued in this book. In fact, if we simply assume a control system of the form

$$\dot{x} = Ax + bu, \quad y = cx,$$

so that

$$y(t) = ce^{At}x_0 + \int_0^t ce^{A(t-s)}bu(s)ds,$$

we are ready to apply a century of results from linear control theory to the problem of smoothing splines. Note, for example, that the choice of

$$A = \begin{pmatrix} 0 & 1 \\ 0 & 0 \end{pmatrix}, \ b = \begin{pmatrix} 0 \\ 1 \end{pmatrix}, \ c = \begin{pmatrix} 1 & 0 \end{pmatrix}$$

corresponds to the situation in (1.2).

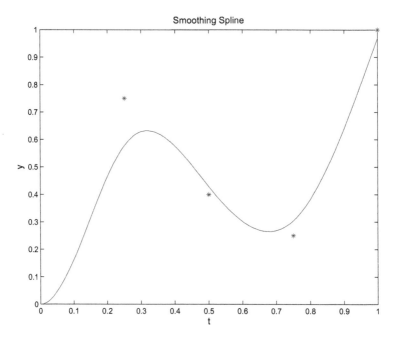

Figure 1.2 Smoothing cubic splines.

1.3.1 When Do Solutions Exist?

It is an easy matter to add additional constraints to the control theoretic formulation of the smoothing spline problem. For example, we can introduce

$$C = \{u \in L_2[0,T] \mid \ell_i(u) = 0, \ i = 1,\ldots,M\},$$

where each ℓ_i is an affine linear functional on $L_2[0,T]$.

We can then ask for the solution to the constrained problem

$$\min_{u \in C} \int_0^T u(t)^2 dt + \lambda \sum_{i=1}^n (y(t_i) - \alpha_i)^2.$$

For example, we might let

$$\ell_1(u) = \int_0^1 (t - s)u(s)ds - 1$$

and

$$\ell_2(u) = \int_0^1 (t - s)u(s)ds + 1.$$

Here there is obviously no solution since the constraints are contradictory. A general condition for the solution to this problem to exist is of course that C is nonempty. In fact, as we see later in the book, C defines an affine subspace in $L_2[0, T]$, and the optimization problem is simply asking for the point of minimum norm in that affine space. *And, as long as the affine space is closed there is guaranteed to be a unique solution, as a direct consequence of Hilbert's famous projection theorem.* (See Section 2.3.) As a consequence, we do not need the full machinery of convex optimization, as developed by Rockafellar [82].

As a final comment, it should be noted that if the constraints are nonlinear, the problem is much more difficult. In fact, even if the constraints define a "nice" subspace of L_2, the problem of constructing the optimal control in this case can be (and usually is) very difficult. We will examine a few such problems in this book.

1.4 APPLICATIONS

One of the major goals of this book is to provide tools for applications. To that end, we consider two main categories of applications to which the control theoretic spline is particularly well suited: *path planning* and *statistics*. In fact, even though the major impetus for this work came from path planning–originally the air traffic control problem–it has evolved into a much more general problem involving many autonomous vehicles or even biological entities.

1.4.1 Path Planning

We consider this problem in several chapters in the book. The basic idea is that we are given a set of way points and times, and we ask that the system be at, or near, those points at specified times. We are not very interested in the nature of the error at the way points unless it is too large. If, for example, we are trying to design a path for an autonomous vehicle, we may have to impose restrictions on the curvature of the path at particular points, and this may require iterations over different choices of smoothing parameters (λ) to deliver a suitable path.

1.4.2 Statistics

The major role of splines in statistics is to smooth noisy data. To this end, it is important that the residues be well behaved. This observation has led to a science studying the selection of the smoothing parameter λ in (1.1) to achieve residues that have suitable statistical properties. Hopefully, λ can be chosen so that the residues are identically normally distributed. This is seldom a goal in engineering. In this book, we use the parameter λ to control bandwidth and do not study the residues as such. However, we will tie into a number of statistically motivated applications, including the production of probability densities using smoothing splines.

1.5 TOPICAL OUTLINE OF THE BOOK

This book is organized into ten chapters (plus the introduction). In Chapter 2, the basic material from control theory is presented as well as the setup for the solution of the optimal control problem. Fundamental concepts from the theory of Hilbert spaces are summarized. Notation is established and, as an example, we revisit the classical controllability problem in the context of Hilbert's projection theorem.

In Chapter 3, we describe eight problems that are fundamental in the area of interpolation and smoothing, and that will serve as motivation for the subsequent chapters. Rather than providing complete solutions to these problems, this chapter should be thought of more as a road-map for this book (and beyond). The eight problems are (1) interpolating splines, (2) interpolating splines with constraints, (3) smoothing splines, (4) smoothing splines with constraints, (5) monotone smoothing splines, (6) dynamic time warping, (7) model following, and (8) trajectory planning. Problems (1) and (3) are basic to the material in the book. The other six problems are important as applications and constitute refinements of the two basic problems.

In Chapter 4, we consider the general problem of smoothing splines from the viewpoint of Hilbert spaces. In some sense, this chapter is the major contribution of control theory techniques to the spline problems, and constitutes the core of the book. We show that the general smoothing splines result from an application of Hilbert's projection theorem, and that we are able to add any finite number of linear constraints to the formulation and still have an effective algorithmic solution.

In Chapter 5, we show that control theoretic splines have the properties that we expect from splines–suitable approximation properties. We show, for example, that if we are given a smooth curve, then, as the number of data points approaches a dense set, the sequence of splines converges in an appropriate manner to this underlying curve. We also show that if noise is added, we still maintain convergence.

In Chapter 6, we consider an extension of the smoothing spline problem with finite/discrete data (a finite/discrete collection of data points) to the problem of smoothing splines with continuous data. This problem is in some real sense a filtering problem. The data can be considered to be the output of some machine, and we are trying to find a smooth approximation of these data. The smoothing spline formulation lends itself well to the problem. In this chapter we also consider the problem of recursive splines as a natural tool for tackling the continuous data problem.

Chapter 7 deals with the question of how to produce splines with certain regularity properties. In particular, we discuss how to produce splines that are monotone in the sense of having nonnegative first or second derivatives. The main theorem in this chapter is that for nilpotent systems, the optimal curve is still piecewise polynomial despite the monotonicity constraints, while the problem is completely solved using dynamic programming for the case of monotone cubic splines. The monotone smoothing problem is of importance in a number of applications ranging from economics to biology. In this chapter we also discuss the related problem of constructing probability density functions from data, which is an example of a much larger problem involving continuous constraints.

In Chapter 8, we further consider the application of smoothing splines to statistics by showing that the smoothing spline can be considered as an approximation to an explicit linear filter. The resulting construction will based on linear-quadratic optimization and its associated theories of Hamiltonians and Riccati transforms.

In Chapter 9, we consider a variation of the smoothing spline problem— transfer between affine varieties. An example where this problem arises is considered in detail (path planning for multi-robot systems), and the problem of transfer between affine varieties is solved in its full generality, although introduced and motivated by this particular example problem. Interestingly, this transfer problem can be considered as a control problem on the manifold of affine subspaces.

In Chapter 10, we consider some applications to path planning. In particular, we study the problem of planning paths for multiple airplanes close to an airport and the problem of reconstructing the paths executed by sea turtles, based on telemetric data. As a consequence of this, we are forced to construct splines on spheres instead of on "flat" Euclidean spaces.

Finally, in Chapter 11, we show that there are classes of problems that do not fall into the Hilbert space setting but are still important and can be solved. The particular application under investigation in this context is the classic problem of selecting appropriate nodal (or data) points. In other words, where do you put the sensors to obtain the information that you need for control? In the context of polynomial interpolation, this problem was of interest a hundred years ago, and it still remains an important and primary problem in certain engineering fields.

Chapter Two

CONTROL SYSTEMS AND MINIMUM NORM PROBLEMS

In this chapter, we establish some basic notation and recall some fundamental results and definitions that will be used throughout the book. In particular, we will discuss linear control systems and Hilbert spaces. The reason for this is that a large portion of the book is dedicated to the problem of generating curves with desirable characteristics and properties, which will be achieved using tools from optimal control. However, rather than using traditional variational methods, we will view a majority of the optimal control problems as affinely constrained, minimum norm problems in appropriate Hilbert spaces.

2.1 LINEAR CONTROL SYSTEMS

Given the state of a linear system (e.g., the position and velocity of a car, the currents in an electric network, or the distribution of susceptible, infected, and immune populations in epidemiology dynamics), denoted by $x(t) \in \mathbb{R}^n$, we will study how this state evolves over time intervals $[0, T]$. Moreover, we will be interested in certain measurable aspects of the system (e.g., distance traveled in the car, voltage across a particular component in the network, or the rate at which healthy individuals become infected), and we denote this measurable output by $y(t) \in \mathbb{R}^p$. The final component needed for understanding the various signals in a linear control system is the control signal, $u(t) \in \mathbb{R}^m$. This is the entity through which the dynamics of the system can be changed (e.g., by stepping on the gas pedal, changing the resistance in a variable resistor, or vaccinating segments of the population.)

2.1.1 State Space Representation

The standard state space representation of a finite-dimensional linear control system is

$$\begin{aligned} \dot{x}(t) &= Ax(t) + Bu(t), \\ y(t) &= Cx(t), \end{aligned} \tag{2.1}$$

where A is an $n \times n$ matrix, B is $n \times m$, and C is $p \times n$. These matrices may or may not be time-varying, but throughout this book we will assume that they are constant matrices. Note also that we will use the (lower-case) notation $\dot{x} = Ax + bu$ and $y = cx$ to signify that $u(t)$ and $y(t)$ are scalars rather than vector.

The solution to (2.1) is given by

$$y(t) = Cx(t) = Ce^{At}x_0 + \int_0^T Ce^{A(t-s)}Bu(s)ds. \tag{2.2}$$

Hence, if we define the function

$$\ell_t(s) = \begin{cases} B^T e^{A^T(t-s)}C^T, & s \le t, \\ 0 & \text{otherwise,} \end{cases} \tag{2.3}$$

as well as

$$\beta_t = e^{A^T t}C^T, \tag{2.4}$$

where the superscript T denotes transpose, we can rewrite (2.2) as

$$y(t) = \beta_t^T x_0 + \int_0^T \ell_t^T(s)u(s)ds. \tag{2.5}$$

Now, as this book focuses on the issue of producing curves that pass through (or close to) given data points $\alpha_1, \alpha_2, \dots, \alpha_N$ at given times $t_1, t_2, \dots, t_N$, where $0 < t_1 < \dots < t_N < T$, we note that

$$y(t_i) = \beta_{t_i}^T x_0 + \int_0^T \ell_{t_i}^T(s)u(s)ds, \tag{2.6}$$

which is to be compared to the data point α_i.

As a final observation, if we stack the ℓ_{t_i} and β_{t_i} together as

$$\ell(s) = \begin{pmatrix} \ell_{t_1}(s) \\ \ell_{t_2}(s) \\ \vdots \\ \ell_{t_N}(s) \end{pmatrix}, \quad \beta = \begin{pmatrix} \beta_{t_1} \\ \beta_{t_2} \\ \vdots \\ \beta_{t_N} \end{pmatrix}, \tag{2.7}$$

we can also define the *Grammians*

$$G = \int_0^T \ell(s)\ell^T(s)ds, \tag{2.8}$$

$$\mathcal{B} = \beta\beta^T, \tag{2.9}$$

which will prove highly useful.

2.1.2 The Basic Problem

Again, assume that we are given a set of data points $\alpha_1, \alpha_2, \ldots, \alpha_N$ and corresponding times $t_1, t_2, \ldots, t_N$. Moreover, assume that the data points are scalars. What we would like to do is to drive the scalar output of a given linear control system "close to" the data points while using as little control energy as possible, under the assumption that the control signal is scalar as well.

If the initial condition is fixed, all we can do is change the control input, and this problem becomes an optimal control problem. The basic formulation is

$$\min_{u \in \mathcal{U}} \rho \int_0^T u^2(s)ds + \sum_{i=1}^N w_i(y(t_i) - \alpha_i)^2, \qquad (2.10)$$

where $\rho > 0$ is the *smoothing parameter* and $w_i > 0$, $i = 1, \ldots, N$, is a weight that determines the relative importance given to the ith data point. Moreover, $\mathcal{U}$ is the space of control signals and $y(t_i) = y_{t_i}$, given in (2.6), will belong to $\mathcal{Y}$, that is, the space of output signals.

If we let

$$\alpha = \begin{pmatrix} \alpha_1 \\ \alpha_2 \\ \vdots \\ \alpha_N \end{pmatrix}, \quad \hat{y} = \begin{pmatrix} y_{t_1} \\ y_{t_2} \\ \vdots \\ y_{t_N} \end{pmatrix}, \quad W = \begin{pmatrix} w_1 & 0 & 0 & \cdots & 0 \\ 0 & w_2 & 0 & \cdots & 0 \\ 0 & 0 & w_2 & \cdots & 0 \\ & & \vdots & \ddots & \vdots \\ 0 & 0 & 0 & \cdots & w_N \end{pmatrix},$$

$$(2.11)$$

then we can rewrite (2.10) as

$$\min_{u \in \mathcal{U}} \rho \int_0^T u^2(s)ds + (\hat{y} - \alpha)^T W(\hat{y} - \alpha). \qquad (2.12)$$

We will solve this and many related problems throughout this book, and already it should be noted that a number of these problems can be cast as minimum norm problems over particular functional spaces. This functional view of linear control systems will prove useful for the developments in this book; to make matters more precise, we first have to establish just what kind of spaces these functional spaces might be.

2.2 HILBERT SPACES

In the previous discussion, we referred to the spaces of input and output signals $\mathcal{U}$ and $\mathcal{Y}$, respectively. In this section, we will discuss some of the properties of a class of such spaces of functions, namely, *Hilbert spaces*.

2.2.1 Vector Spaces

A vector space is a set $\mathcal{X}$ on which the two operations of addition and scalar multiplication have been defined. In particular, $\omega_1 + \omega_2 \in \mathcal{X}$ for any two elements $\omega_1, \omega_2 \in \mathcal{X}$, while $\alpha\omega \in \mathcal{X}$ for any $\alpha \in \mathbb{R}$ and $\omega \in \mathcal{X}$. These two operations, moreover, satisfy the following axioms:

(i)	$\omega_1 + \omega_2 = \omega_2 + \omega_1$	(Commutative Law)
(ii)	$(\omega_1 + \omega_2) + \omega_3 = \omega_1 + (\omega_2 + \omega_3)$	(Associative Law)
(iii)	$\exists 0_{\mathcal{X}} \in \mathcal{X}$ s.t. $\omega + 0_{\mathcal{X}} = \omega, \ \forall \omega \in \mathcal{X}$	(Null Element)
(iv)	$\alpha(\omega_1 + \omega_2) = \alpha\omega_1 + \alpha\omega_2$	
(v)	$(\alpha + \beta)\omega = \alpha\omega + \beta\omega$	(Distributive Law)
(vi)	$(\alpha\beta)\omega = \alpha(\beta\omega)$	(Associative Law)
(vii)	$0\omega = 0_{\mathcal{X}}, \ 1\omega = \omega$	

Since a key issue in this book is to be able to chose the "best" control input that makes a linear control system behave in a prescribed manner, some notion of what "best" means is needed. And, in particular, we need to define a concept of *distance* in a vector space. In fact, a *normed, linear vector space* is a vector space $\mathcal{X}$ associated with a norm, satisfying

(i)	$\|\omega\| \geq 0, \ \forall \omega \in \mathcal{X}, \quad \|\omega\| = 0 \Leftrightarrow \omega = 0_{\mathcal{X}}$	
(ii)	$\|\omega_1 + \omega_2\| \leq \|\omega_1\| + \|\omega_2\|, \ \forall \omega_1, \omega_2 \in \mathcal{X}$	(Triangle Inequality)
(iii)	$\|\alpha\omega\| = \|\alpha\| \cdot \|\omega\|, \ \forall \alpha \in \mathbb{R}, \ \omega \in \mathcal{X}.$	

Equipped with a norm, one can define concepts like convergence of sequences $\{\omega_m\}$. A particular type of sequence is the Cauchy sequence, which satisfies $\|\omega_m - \omega_{m'}\| \to 0$ as $m, m' \to \infty$. $\mathcal{X}$ is said to be *complete* if every Cauchy sequence made up of elements in $\mathcal{X}$ also has a limit that remains in $\mathcal{X}$ itself. A complete, normed, linear vector space is called a *Banach space*. However, some additional structure is needed, in particular, the important notion of an *inner product*.

2.2.2 Inner Products

In order to solve a number of the optimal control problems under consideration in this book, we need a notion of orthogonality. Orthogonality is defined through the inner product $\langle \omega_1, \omega_2 \rangle$, with the induced norm $\|\omega\| = \sqrt{\langle \omega, \omega \rangle}$.

That this is indeed a norm must of course be checked against the axioms in the previous section. For example, the properties that

(i) $\langle \omega_1 + \omega_2, \omega_3 \rangle = \langle \omega_1, \omega_3 \rangle + \langle \omega_2, \omega_3 \rangle,$
(ii) $\langle \alpha \omega_1, \omega_2 \rangle = \alpha \langle \omega_1, \omega_2 \rangle,$

directly establishes the triangle inequality as follows:

$$\begin{aligned}
\|\omega_1 + \omega_2\|^2 &= \langle \omega_1 + \omega_2, \omega_1 + \omega_2 \rangle \\
&= \langle \omega_1, \omega_1 \rangle + \langle \omega_1, \omega_2 \rangle + \langle \omega_2, \omega_1 \rangle + \langle \omega_2, \omega_2 \rangle \\
&\leq \|\omega_1\|^2 + 2|\langle \omega_1, \omega_2 \rangle| + \|\omega_2\|^2.
\end{aligned}$$

Now, the *Cauchy-Schwarz inequality* states that

$$|\langle \omega_1, \omega_2 \rangle| \leq \|\omega_1\| \cdot \|\omega_2\|, \; \forall \omega_1, \omega_2 \in \mathcal{X};$$

hence we have that

$$\begin{aligned}
\|\omega_1 + \omega_2\|^2 &\leq \|\omega_1\|^2 + 2\|\omega_1\| \cdot \|\omega_2\| + \|\omega_2\|^2 \\
&= (\|\omega_1\| + \|\omega_2\|)^2,
\end{aligned}$$

and the triangle inequality follows.

Now, the final construction needed to be able to properly define what we mean by orthogonality and projections is the notion of a *Hilbert space*. And, a Hilbert space is simply a Banach space with an inner product that induces the norm.

2.3 THE PROJECTION THEOREM

Assume that $\mathcal{H}$ is a Hilbert space and that V is a closed subspace of $\mathcal{H}$. Given an arbitrary point $p \in \mathcal{H}$, a classic problem is that of trying to find the point in V that is closest to p. In the finite-dimensional case, we know that this point is given by the projection of p onto V. One remarkably powerful fact about Hilbert spaces is that, in this regard, they behave just like finite-dimensional spaces. It is, in fact, possible to talk about projections in Hilbert spaces in a straightforward manner, through the notion of *orthogonality*.

Two elements $p, p' \in \mathcal{H}$ are said to be orthogonal if $\langle p, p' \rangle = 0$, denoted by $p \perp p'$. (As an example, an immediate consequence of orthogonality is that if $p \perp p'$ then $\|p + p'\|^2 = \|p\|^2 + \|p'\|^2$.) We say that $p \perp S$ if $p \perp s$, $\forall s \in S$, where S is any subset of $\mathcal{H}$.

Armed with the notion of orthogonality, we can thus state Hilbert's famed projection theorem:

Theorem 2.1 (Hilbert's projection theorem) *Let p be a point in a Hilbert space $\mathcal{H}$ together with a closed subspace V of $\mathcal{H}$. There exists a unique point $v_0 \in V$ that is the closest to p in the sense that $\|p - v_0\| \leq \|p - v\|$, $\forall v \in V$. Moreover, v_0 is uniquely determined by the condition that $x - v_0 \perp V$.*

For example, if we let $F : \mathcal{H} \to \mathbb{R}^m$ be a linear operator that maps points in $\mathcal{H}$ to $\mathbb{R}^m$, we can define the subspace V as

$$V = \{w \in \mathcal{H} \mid Fw = 0\}.$$

Moreover, if we let $r \in \mathbb{R}^m$, we can translate V to get the *affine variety* V_r, given by

$$V_r = \{w \in \mathcal{H} \mid Fw = r\},$$

where we note that $V_0 = V$.

As before, let $p \in \mathcal{H}$ be an arbitrary point in $\mathcal{H}$, and consider the minimum norm problem

$$\min_{w \in \mathcal{H}} \|w - p\|^2 \tag{2.13}$$
$$\text{such that } w \in V_r.$$

This problem is depicted in Figure 2.1, where it is also shown how the projection theorem gives the unique minimizer.

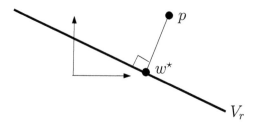

Figure 2.1 Solution to the problem in (2.13).

2.3.1 Finding the Minimizer

Unfortunately, it does not follow that just because it is clear pictorially, it is clear how to construct the unique solution to the problem in (2.13). Following the development in [64], we will describe the solution method that will be used repeatedly throughout this book.

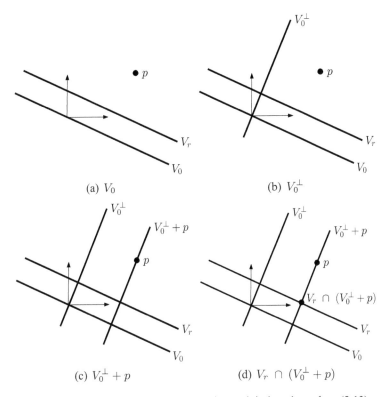

(a) V_0 (b) $V_0^\perp$

(c) $V_0^\perp + p$ (d) $V_r \cap (V_0^\perp + p)$

Figure 2.2 The steps to computing the unique minimizer that solves (2.13).

The first part in solving the problem in (2.13) is to translate V_r back to a subspace, that is, to find V_0, as shown in Figure 2.2(a). Since this is a subspace, it is easy to compute its orthogonal complement $V_0^\perp$, given by

$$V_0^\perp = \{w \in \mathcal{H} \mid w \perp V_0\},$$

as shown in Figure 2.2(b).

The next step toward finding the unique minimizer is to translate the orthogonal complement by p to obtain $V_0^\perp + p$, as

$$V_0^\perp + p = \{w \in \mathcal{H} \mid w = \omega + p, \text{ for some } \omega \in V_0^\perp\}.$$

This step is shown in Figure 2.2(c), and the unique minimizer $w^\star$ is now directly found by computing the intersection

$$\{w^\star\} = V_r \cap (V_0^\perp + p),$$

as in Figure 2.2(d).

We summarize these steps below:

Given

- $\mathcal{H}$ a Hilbert space,

- $V_r \subset \mathcal{H}$ an affine variety in $\mathcal{H}$,

- $p \in \mathcal{H}$,

- the minimization problem

$$\min_{w \in \mathcal{H}} \|w - p\|^2$$
$$\text{such that } w \in V_r,$$

the unique minimizer $w^\star$ is given by the projection of p onto V_r, computed through:

1. Find V_0.

2. Compute the orthogonal complement $V_0^\perp$.

3. Translate to $V_0^\perp + p$.

4. $w^\star$ is given uniquely by the intersection $V_r \cap (V_0^\perp + p)$.

It should noted, at this point that, even though these four steps needed to solve the problem in (2.13) may seem simple, they are in fact rather powerful tools that can be used in a number of problems involving smoothing splines. This will be demonstrated in subsequent chapters.

2.4 OPTIMIZATION AND GATEAUX DERIVATIVES

Most of the problems we will encounter in this book can be reformulated as minimum norm problems in Hilbert spaces. The constraints will mostly take on the form of memberships in certain affine varieties. Thus, the projection theorem, as discussed in Section 2.3, is directly applicable. However, it will not always be the case that we can apply this method, which calls for slightly more complex machinery. Even though the basic smoothing problem, as discussed in Section 2.1.2, can be cast as a minimum norm problem in a Hilbert space, variations to this problem cannot. Problems that cannot be solved using the projection theorem include

- The Sample-Point Selection Problem. Given that the data come from an underlying generator $h(t)$, how can we pick the best times $t_1, \ldots, t_N$ so that the corresponding data points $\alpha_1 = h(t_1), \ldots, \alpha_N = h(t_N)$ are optimal, in a certain sense?

- The Monotone Smoothing Problem. Given the basic problem formulation in Section 2.1.2, can we solve this problem subject to monotonicity constraints on the output y, or on its derivatives?

In order to address these and related problems, we will need to employ other techniques for solving the optimal control problems.

2.4.1 Parameter Optimization

Consider the problem of finding the minimizer $\xi^* \in \mathbb{R}^m$ to the function $h : \mathbb{R}^m \rightarrow \mathbb{R}$, under the constraint that $\xi \in S \subset \mathbb{R}^m$. The optimality conditions to this problem typically depend on the constraint set S, and if S is given by equality constraints $S = \{\xi \mid G(\xi) = 0\}$, with $G : \mathbb{R}^m \rightarrow \mathbb{R}^p$, under certain regularity conditions on h and G, we get the well-known first-order necessary optimality conditions.

Equality Constraints
Let ξ^* be a (regular) local extremum to $h(\xi)$ under the constraint that $G(\xi) = 0 \in \mathbb{R}^p$. Then there exists a Lagrange multiplier $\lambda \in \mathbb{R}^p$ such that

$$\frac{\partial h(\xi^*)}{\partial \xi} + \lambda^T \frac{\partial G(\xi^*)}{\partial \xi} = 0.$$

The corresponding first-order necessary conditions (the so-called *Kuhn-Tucker conditions*), when the constraint set is given by inequality conditions $G(\xi) \leq 0$, are as follows:

Inequality Constraints
Let ξ^* be a (regular) local extremum to $h(\xi)$ under the constraint that $G(\xi) \leq 0 \in \mathbb{R}^p$. Then there exists a Lagrange multiplier $\lambda \in \mathbb{R}^p$ such that

$$\frac{\partial h(\xi^\star)}{\partial \xi} + \lambda^T \frac{\partial G(\xi^\star)}{\partial \xi} = 0,$$

$$\lambda^T G(\xi^\star) = 0,$$

$$\lambda \geq 0.$$

Even though these conditions are standard fare in any book on optimization, we choose to include them for the sake of easy reference.

2.4.2 Optimization in Functional Spaces

Similarly, we can treat optimization problems on functional spaces as problems involving stationary points. However, it is not at all clear what the operation of $\partial/\partial\xi$ actually corresponds to in these spaces. In fact, as normal derivatives can be thought of as limiting concepts, one can ask how much a given functional $F : \mathcal{H} \to \mathbb{R}$ increases at a certain point $p \in \mathcal{H}$ if we allow a small perturbation of p. In the functional case, this limiting increment is not independent of the perturbation, and we will have to talk about *directional derivatives* rather than normal derivatives.

Let, as before, $\mathcal{H}$ be a Hilbert space and let $F : \mathcal{H} \to \mathbb{R}$. By the *Gateaux differential* of F at p along q, we understand

$$\delta F(p, q) = \lim_{\epsilon \to 0} \frac{F(p + \epsilon q) - F(\epsilon)}{\epsilon}. \tag{2.14}$$

Analogous to the unconstrained, finite-parameter optimization situation, a necessary optimality condition is that the Gateaux differential vanishes along all directions q.

Unconstrained Case
Let $p^\star$ be a local extremum to $F : \mathcal{H} \to \mathbb{R}$. Then $\delta F(p^\star, q) = 0$, $\forall q \in \mathcal{H}$.

The constrained cases are significantly more involved, but we can, for example, obtain the classic optimality conditions in optimal control from the computation of the Gateaux differential. The way in which these can be obtained is to insist that the control signal u belongs to a Hilbert space, $\mathcal{U}$, and to view the dynamics as an equality constraint defined for all times. Then the variation can be computed along the direction v, which is the same as computing the Gateaux differential at u along v. Ensuring that this differential vanishes for all directions then produces the classic first-order necessary optimality conditions in variational calculus. (See, e.g., [10].) The Gateaux derivative will be put to heavy use in this book in general, and Chapter 5 in particular.

2.5 THE POINT-TO-POINT TRANSFER PROBLEM

2.5.1 Control Systems as Linear Operators

For the purpose of establishing and enforcing the fact that linear control systems may be thought of as mapping control signals, defined over a suitable functional space, into the state (or output) space, we make use of the classic point-to-point transfer problem. In fact, this problem is particularly well suited for this purpose in that it explicitly involves conditions for controllability.

As before, we assume that the linear dynamics is given by the following time-invariant system

$$\dot{x}(t) = Ax(t) + Bu(t), \tag{2.15}$$

where $x(t) \in \mathbb{R}^n$ and $u(t) \in \mathbb{R}^m$. However, rather than stressing the fact that u at a particular time is a vector in $\mathbb{R}^m$, one can view u itself as a point in a functional space. For example, we can insist that $u \in L_2^m[0, T]$, which is the (Hilbert) space of equivalent classes of square-integrable m-dimensional functions. (We will often suppress the dependence on m and $[0, T]$ and simply write L_2 whenever it is clear from the context what space is considered.) With a slight abuse of notation (ignoring the equivalence class issue), we can think of $L_2^m[0, T]$ as

$$L_2^m[0, T] = \left\{ w : [0, T] \rightarrow \mathbb{R}^m \text{ such that } \int_0^T w^T(t)w(t)dt < \infty \right\},$$

with inner product

$$\langle v, w \rangle_{L_2} = \int_0^T v^T(t)w(t)dt.$$

Now, the solution to (2.15), given that the initial state $x(0) = x_0$, is given by

$$x(T) = e^{AT}x_0 + \int_0^T e^{A(T-t)}Bu(t)dt. \tag{2.16}$$

And, if we define the linear operator $\Lambda : L_2 \rightarrow \mathbb{R}^n$ as

$$\Lambda u = \int_0^T e^{A(T-t)}Bu(t)dt,$$

the solution in (2.16) can be reformulated as

$$x(T) = e^{AT}x_0 + \Lambda u. \tag{2.17}$$

2.5.2 Controllability and the Point-to-Point Transfer Problem

The point-to-point transfer problem considers whether it is possible to drive x from $x(0) = x_0$ to a given $x(T) = x_T$. In light of (2.17), the point-to-point transfer problem is exactly that of determining if there exists a u such that

$$x_T - e^{AT} x_0 = \Lambda u,$$

or equivalently if

$$x_T - e^{AT} x_0 \in \mathcal{R}(\Lambda),$$

where the range space of Λ is given by

$$\mathcal{R}(\Lambda) = \{z \in \mathbb{R}^n \mid \exists u \in L_2 \text{ such that } z = \Lambda u\}.$$

Since L_2 is an infinite-dimensional vector space, it may not be a particularly easy task to characterize this range space. However, we know that

$$\mathcal{R}(\Lambda) = \mathcal{R}(\Lambda \Lambda^\star),$$

where the adjoint operator $\Lambda^\star : \mathbb{R}^n \to L_2$ is defined through

$$\langle z, \Lambda v \rangle_{\mathbb{R}^n} = \langle \Lambda^\star z, v \rangle_{L_2}.$$

Note that $\Lambda \Lambda^\star : \mathbb{R}^n \to \mathbb{R}^n$ is simply an $n \times n$ matrix, which means that $\mathcal{R}(\Lambda \Lambda^\star)$ should be easily computed.

What remains is to compute the adjoint operator. We have

$$\langle z, \Lambda v \rangle_{\mathbb{R}^n} = z^T \int_0^T e^{A(T-t)} B u(t) dt = \int_0^T \left(B^T e^{A^T(T-t)} z \right)^T u(t) dt,$$

which means that

$$[\Lambda^\star z](t) = B^T e^{A^T(T-t)} z,$$

and

$$\Lambda \Lambda^\star = \int_0^T e^{A(T-t)} B B^T e^{A^T(T-t)} dt.$$

This is the controllability Grammian, denoted by Γ, and we have thus derived the classic result that the point-to-point transfer problem has a solution if and only if

$$x_T - e^{AT} x_0 \in \mathcal{R}(\Gamma).$$

As a special case, we can consider the situation where

$$\mathrm{rank}\,(\Gamma) = n.$$

In this case, it is possible to drive between any initial and final states, and if the controllability Grammian satisfies this property, we say that the system is completely controllable.

2.5.3 Minimum Energy Transfer

Now that we know when the point-to-point transfer problem has a solution, a reasonable question is to try to solve it using as little control effort as possible. In fact, we would like to solve it while minimizing $\|u\|_{L_2}^2$. And, if we let

$$\rho = x_T - e^{AT}x_0,$$

we can define the affine variety

$$V_\rho = \{v \in L_2 \mid \rho = \Lambda v\}.$$

Based on this notation, we have arrived at a formulation that renders the projection theorem applicable, namely,

$$\min_{u \in L_2} \|u\|_{L_2}^2$$
$$\text{such that } u \in V_\rho.$$

Following the discussion about the projection theorem in Section 2.3, we know that the unique minimizer is given by

$$\{u^\star\} = V_0^\perp \cap V_\rho,$$

where $V_0 = \{v \in L_2 \mid 0 = \Lambda v\}$ is exactly the null space of Λ, $\mathcal{N}(\Lambda)$. A fundamental fact about null and range spaces is that

$$\mathcal{N}(\Lambda) = \mathcal{R}(\Lambda^\star)^\perp,$$

and hence

$$V_0^\perp = \mathcal{R}(\Lambda^\star) = \{v \in L_2 \mid \exists z \in \mathbb{R}^n \text{ such that } \Lambda^\star z = v\}.$$

All that remains in order to find the unique minimizer is to compute the intersection $V_0^\perp \cap V_\rho$. But, $v \in V_0^\perp \Leftrightarrow \Lambda^\star z = v$ for some $z \in \mathbb{R}^n \Leftrightarrow \Lambda\Lambda^\star z = \Lambda v$ for some $z \in \mathbb{R}^n$.
  We know that $\rho = \Lambda v$ since $v \in V_\rho$, and hence

$$\Lambda\Lambda^\star z = \rho.$$

Now, if the system is assumed to be completely controllable, we have that

$$z = (\Lambda\Lambda^\star)^{-1}\rho,$$

which in turn gives the unique minimizer as

$$u^\star = \Lambda^\star(\Lambda\Lambda^\star)^{-1}\rho.$$

As a result, we have solved an optimal control problem in a purely geometric fashion. If we use the definitions of Γ, $\Lambda^\star$, and ρ, we arrive at the familiar control law

$$u^\star(t) = B^T e^{A^T(T-t)}\Gamma^{-1}(x_T - e^{AT}x_0).$$

In a similar fashion, we will be able to solve more and more involved optimal control problems that arise in the control theory literature, as well as in statistics and numerical analysis.

Chapter Three

EIGHT FUNDAMENTAL PROBLEMS

In this chapter, we introduce a series of eight fundamental problems in the areas of interpolation and smoothing of increasing complexity. These problems will serve as basic building blocks for the developments in later chapters; in particular, we show that, although these eight problems have their origins in optimal control theory, statistics, and numerical analysis, they can be addressed in a unified manner. It will be shown that splines and linear optimal control theory are very closely related to the classical numerical theory of splines. We will also show that all eight problems are very similar in nature, if not in solution. We will not, at this point, effectively solve all eight problems, but rather hint as to the solutions that will be derived in later chapters.

In Problem 1, we show that the theory of interpolating splines is naturally considered as a problem of minimizing a quadratic cost functional subject to a set of linear constraints, and in Problem 3, we show that the theory of smoothing splines can be considered as very close to the theory of interpolating splines, the difference being that the linear constraints are included in the cost functional as a penalty term. For both Problems 1 and 3, the optimization problem is a straightforward problem of minimizing a quadratic cost functional over the space of square integrable functions.

In Problems 2 and 4, we show that the problem of constructing splines that pass through intervals instead of points can be reduced to the problem of minimizing a quadratic cost subject to a set of inequality constraints. While it is known that not all quadratic programming problems have solutions, this has not been an issue with problems associated with splines.

In Problem 5, we discuss the problem of constructing splines that are nondecreasing at the nodes. This will, following the same line of reasoning as for the previous problems, be reducible to the problem of minimizing a quadratic cost functional with inequality constraints, and hence reducible to a quadratic programming problem.

In Problem 6, we restate the curve registration problem of Li and Ramsay [58] as an optimal output tracking problem. Here, the problem becomes noticeably harder since the cost functional as well as the differential constraints are now nonlinear. In Problem 7, we state the general output tracking problem and show that Problem 6 is indeed a special case of this problem.

Finally, in Problem 8, we state a version of the trajectory planning problem. The statement of this problem involves the previous seven problems. Although the solution is not given, we do present an algorithm that will at least produce a suboptimal solution. It should in fact be stressed that none of these problems will be solved to completion in this chapter, but rather they are to be thought of as motivating the further developments in later chapters as well as future research.

3.1 THE BASIC SET-UP

3.1.1 Assumptions on the Underlying System Dynamics

Following the notation in Chapter 2, we will assume that we are given a linear, time-invariant dynamical system of the form

$$\dot{x}(t) = Ax(t) + bu(t), \tag{3.1}$$
$$y(t) = cx(t), \tag{3.2}$$

where $x(t) \in \mathbb{R}^n$. Note that we assume that for this system $u(t)$ and $y(t)$ are scalars.

We will, furthermore, sometimes find it useful to impose some constraints on the relative degree of the system through the assumption that

$$cb = cAb = cA^2b = \cdots = cA^{n-2}b = 0. \tag{3.3}$$

This condition can be relaxed, but parts of the exposition are simplified and, as we will see, maximal smoothness is obtained with this assumption.

A canonical example of a system that satisfies the relative degree assumption is

$$A = \begin{pmatrix} 0 & 1 & 0 & \cdots & 0 \\ 0 & 0 & 1 & \cdots & 0 \\ \vdots & & & \ddots & \vdots \\ 0 & 0 & 0 & \cdots & 1 \\ \alpha_1 & \alpha_2 & \alpha_3 & \cdots & \alpha_n \end{pmatrix}, \tag{3.4}$$

$$b = \begin{pmatrix} 0 \\ 0 \\ \vdots \\ 0 \\ 1 \end{pmatrix}, \quad c = \begin{pmatrix} 1 & 0 & \cdots & 0 & 0 \end{pmatrix}.$$

In fact, any system which satisfies the relative degree constraints in (3.3) is equivalent to a system on this form. For example, consider the problem of controlling a linear spring connected to a unit mass. If we let y be the position of the mass, and let u be an externally applied force, then Newton's Second Law dictates that

$$\ddot{y}(t) = -\delta\dot{y}(t) - ky(t) + u(t),$$

where k is the spring coefficient and δ is the damping coefficient.

Now, setting $x_1 = y$, $x_2 = \dot{y}$, and $x = (x_1, x_2)^T$, we get

$$\dot{x}(t) = \begin{pmatrix} 0 & 1 \\ -k & -\delta \end{pmatrix} x(t) + \begin{pmatrix} 0 \\ 1 \end{pmatrix} u(t),$$
$$y(t) = (\ 1 \quad 0\ ) x(t),$$

which is of the prescribed form.

The case in which all the α_i in (3.4) are zero plays a particularly important role, for then the solutions to the differential equation are polynomials convolved with the control function u. Under this assumption, all that follows in this chapter reduces to the case of polynomial splines.

3.1.2 The Data Sets

The data sets we consider in this chapter are of two basic types, namely *deterministic* and *random*. For trajectory planning problems, we usually consider the data to be given in a deterministic form, that is, the coordinates of the locations and times are given exactly. We denote this "deterministic data" set by

$$DD = \{(\alpha_i, t_i),\ i = 1, \ldots, N \mid 0 < t_1 < t_2 < \cdots < t_N < T\}, \quad (3.5)$$

where DD stands for deterministic data, and where we are interested in the behavior of the system over the time interval $[0, T]$.

As an example, consider the problem of air traffic control, where high-level paths are given in terms of target locations together with the times at which the aircraft is supposed to be at the different locations. The problem to be solved in this situation is to plan paths that take the aircraft close to the target locations at the specified times.

In contrast to this, consider the situation in which the data points are obtained from noisy measurements of an underlying curve $f : [0, T] \to \mathbb{R}$. When addressing the problem of reconstructing f, one is faced with the issue of handling stochastic rather than deterministic data. We denote by

$$SD = \{(f(t_i) + \epsilon_i, t_i),\ i = 1, \ldots, N \mid 0 < t_1 < t_2 < \cdots < t_N < T\}$$
$$(3.6)$$

this stochastic data set, where the ϵ_i are observed values of a random variable which, in general, we assume is symmetric with mean 0. The term SD is to be interpreted as "stochastic data," and this data set SD is the set usually encountered in statistics.

3.1.3 Solving the Linear System

Solving the differential equation in (3.1), (3.2), we have

$$y(t) = ce^{At}x_0 + \int_0^t ce^{A(t-s)}bu(s)ds. \tag{3.7}$$

It is convenient to set $x_0 = 0$ since the initial data can be absorbed into the data set, as will be shown in later chapters.

We now rewrite this solution based on the one-parameter family of functions in (2.3) from the previous chapter:

$$\ell_t(s) = \begin{cases} ce^{A(t-s)}b, & t > s, \\ 0 & \text{otherwise}, \end{cases} \tag{3.8}$$

where we have changed the notation slightly from (2.3). However, since $\ell_t(s)$ is scalar, the two formulations are equivalent.

We can now define an output version of the linear operator from the previous chapter,

$$L_t(u) = \int_0^T \ell_t(s)u(s)ds, \tag{3.9}$$

which, together with the assumption that $x(0) = 0$, gives us the fundamental relationship

$$y(t) = L_t(u). \tag{3.10}$$

We will, moreover, make use of a differential version of this functional, namely,

$$D^k L_t(u) = \int_0^T \frac{d^k \ell_t}{dt^k}(s)u(s)ds, \tag{3.11}$$

and we note that this derivative is well defined, provided that $k < n - 2$, under the assumption of (3.3). Using this notation, we have

$$\frac{d^k}{dt^k}y(t) = D^k L_t(u). \tag{3.12}$$

3.2 INTERPOLATING SPLINES

In this section, we consider the first fundamental problem, namely, the problem of constructing a control law $u(t)$ that drives the output function $y(t)$ through a set of data points at prescribed times. We will construct u so that the resulting output curve is piecewise smooth and generalizes the classical concept of polynomial splines. For this, we will consider the data set DD.

The interpolating conditions can be expressed as

$$\alpha_i = y(t_i) = L_{t_i}(u), \ i = 1, \ldots, N. \tag{3.13}$$

There are, of course, infinitely many control laws that satisfy these constraints. The problem is to identify a scheme that will select a unique control law in some meaningful way. As already mentioned, linear quadratic optimal control provides a convenient tool for this selection and, in fact, the main objective of this book is to show that optimal control plays a natural role for this.

For the sake of keeping things simple, we will here consider the energy cost functional

$$J(u) = \int_0^T u^2(s)ds. \tag{3.14}$$

It should be noted that it is possible to increase the complexity of the cost functional, which is the case in references [4],[5].

3.2.1 Problem 1

As seen in the previous chapter, in order for the optimal control problem to be well posed, we must specify from what set the control is to be chosen, and here we insist that $u \in L_2[0, T]$. The first of the fundamental problems (the interpolation problem) then becomes

---

Problem 1: Interpolating Splines

$$\min_{u \in L_2} J(u)$$

subject to the N constraints

$$\alpha_i = L_{t_i}(u), \ i = 1, \ldots, N.$$

This problem can be easily solved using the techniques discussed in the previous chapter. In fact, the affine variety in $L_2[0, T]$ that we are interested in is defined through the constraints

$$V_\alpha = \{u \in L_2 \mid \alpha_i = L_{t_i}(u), \ i = 1, \ldots, N\}.$$

Following the procedure from the projection theorem, we first have to construct the orthogonal complement to the linear subspace defined by

$$V_0 = \{u \in L_2[0, T] \mid L_{t_i}(u) = 0, \ i = 1, \ldots, N\}.$$

It is straightforward to see that this set is the same as the set spanned by the functions $\ell_{t_i}(s)$. Thus, the optimal control is of the form

$$u^\star(s) = \sum_{i=1}^{N} \tau_i \ell_{t_i}(s), \tag{3.15}$$

for some scalars $\tau_1, \ldots, \tau_N$. What is remarkable about this is that the optimal control problem, which is an inherently infinite-dimensional problem, has been transformed into a finite-dimensional problem involving finding the parameters $\tau_1, \ldots, \tau_N$. In statistics, Problem 1 would be referred to as a nonparametric problem but, due to its solution being reducible to (3.15), it is in fact a semiparametric problem.

Substituting (3.15) into the equations defining the affine variety, we have a set of equations

$$y(t_1) = \tau_1 L_{t_1}(\ell_{t_1}) + \cdots + \tau_N L_{t_1}(\ell_{t_N}),$$
$$\vdots$$
$$y(t_N) = \tau_1 L_{t_N}(\ell_{t_1}) + \cdots + \tau_N L_{t_N}(\ell_{t_N}).$$

As in Chapter 2, we now let $\hat{y} = (y(t_1), \ldots, y(t_N))^T$, $\alpha = (\alpha_1, \ldots, \alpha_N)^T$, and $\tau = (\tau_1, \ldots, \tau_N)^T$, which allows us to write the previous set of linear equations in matrix form as

$$\hat{y} = G\tau = \alpha \ \Leftrightarrow \ \tau = G^{-1}\alpha,$$

where G is the positive definite Grammian

$$G = \begin{pmatrix} L_{t_1}(\ell_{t_1}) & \cdots & L_{t_1}(\ell_{t_N}) \\ \vdots & & \vdots \\ L_{t_N}(\ell_{t_1}) & \cdots & L_{t_N}(\ell_{t_N}) \end{pmatrix} = \int_0^T \ell(s)\ell^T(s)ds,$$

with $\ell(s) = (\ell_{t_1}(s), \ell_{t_2}(s), \ldots, \ell_{t_N}(s))^T$.

Now, in light of (3.15), we have that

$$u^\star(t) = \tau^T \ell(t);$$

that is, the optimal solution becomes

$$u^\star(t) = \alpha^T G^{-1} \ell(t).$$

It should be noted that since the matrix G is in fact a Grammian, it has the potential to be poorly conditioned. However, the advantage of this formulation is that it is immediately clear that there is a unique solution since the $\ell_{t_i}(s)$ are linearly independent. The conditioning can be greatly improved by replacing the functions $\ell_{t_i}(s)$ with a set of functions that are nonzero only on intervals of the form $[t_i, t_{i+n}]$. This procedure is outlined in [58]. The advantage of this is that it reduces the matrix G to a banded matrix (tridiagonal in the case of $n = 2$), which somewhat simplifies the solution.

There are several ways to construct splines to solve the basic problem. A totally different construction that is much better conditioned is developed in [100]. That construction develops the banded structure directly, but has the disadvantage of not carrying over to the more general problems that are pursued in this book.

3.2.2 Example

As an example, consider the classic cubic, interpolating splines, where $\ddot{y}(t) = u(t)$, that is, where

$$A = \begin{pmatrix} 0 & 1 \\ 0 & 0 \end{pmatrix}, \quad b = \begin{pmatrix} 0 \\ 1 \end{pmatrix}, \quad c = \begin{pmatrix} 1 & 0 \end{pmatrix}.$$

An example of solving Problem 1 for this system is shown in Figure 1.1, where the paramaters used are

$$T = 1, \ N = 4,$$
$$t_1 = 1/4, \ t_2 = 1/2, \ t_3 = 3/4, \ t_4 = 1,$$
$$\alpha_1 = 3/4, \ \alpha_2 = 2/5, \ \alpha_3 = 1/4, \ \alpha_4 = 1.$$

3.3 INTERPOLATING SPLINES WITH CONSTRAINTS

In this section, we consider a somewhat different problem that arises in a number of applications and that can be solved in much the same manner as for the classical interpolating spline. The problem we consider is when the system must be driven through intervals instead of data points.

3.3.1 Problem 2

Problem 2: Interpolating Splines with Constraints

$$\min_{u \in L_2} J(u)$$

subject to the constraints

$$a_i \leq L_{t_i}(u) \leq b_i, \ i = 1, \ldots, N.$$

This problem is discussed in the survey by Wegman and Wright [97]. In this section, we show that this type of spline can indeed be recovered using standard optimal control techniques similar to those in the last section, taken together with some basic tools from mathematical programming. This is needed since we no longer have an affine variety in a Hilbert space, and so, are forced to use somewhat more elaborate tools.

We first note that, because of linearity, the set of controls that satisfy the constraints is closed and convex.

Lemma 3.1 *The set of controls that satisfy the constraints of Problem 2 is a closed and convex subset of $L_2[0, T]$.*

Proof. Variations of the proof can be found in any number of textbooks. Suppose that, for some finite number M, we have controls $u_k(s)$, $k = 1, \ldots, M$, which satisfy the constraints of Problem 2. Then, for each i, we have

$$a_i = \sum_{k=1}^{M} \sigma_k a_i \leq \sum_{k=1}^{M} \sigma_k L_{t_i}(u_k) \leq \sum_{k=1}^{M} \sigma_k b_i = b_i,$$

where

$$\sum_{k=1}^{M} \sigma_k = 1, \ \sigma_k > 0, \ k = 1, \ldots, M.$$

Now consider

$$\sum_{k=1}^{M} \sigma_k L_{t_i}(u_k) = L_{t_i} \left(\sum_{k=1}^{M} \sigma_k u_k \right).$$

Thus, the convex sum of controls satisfies the constraints if the individual controls satisfy the constraints. On the other hand, assume that $\{u_k\}$ is a sequence of controls that each satisfy the constraints. Passing the limit through the integral, because of the compactness of the interval $[0, T]$, it follows that the limit also satisfies the constraints. The lemma follows. ∎

The existence and uniqueness of the optimal control follows from standard theorems, and we state this without proof. (See, for example, [96].)

Theorem 3.2 *There exists a unique control signal $u(t)$ that satisfies the constraints of Problem 2 and that minimizes $J(u)$.*

Now, in order to solve Problem 2, we first let

$$a = (a_1, a_2, \ldots, a_N)^T$$

and

$$b = (b_1, b_2, \ldots, b_N)^T.$$

Since Problem 2 is a constrained optimal control problem, we introduce the Lagrange multipliers λ, γ. In order to find the unique solution satisfying the Kuhn-Tucker first-order necessary optimality conditions, we form the associated optimal control problem

$$\max_{\gamma \geq 0, \lambda \geq 0} \min_{u \in L_2} H(u, \lambda, \gamma), \tag{3.16}$$

where the positivity constraints over λ and γ are taken component-wise, and where

$$H(u, \lambda, \gamma) = \frac{1}{2} \int_0^T u^2(t)dt + \sum_{i=1}^N \lambda_i(a_i - L_{t_i}(u)) + \sum_{i=1}^N \gamma_i(L_{t_i}(u) - b_i) \tag{3.17}$$

and

$$\gamma^T = (\gamma_1, \gamma_2, \ldots, \gamma_N)^T, \quad \lambda^T = (\lambda_1, \lambda_2, \ldots, \lambda_N)^T.$$

We first minimize the function H over u, assuming that λ and γ are fixed. This minimum is achieved at the point where the Gateaux derivative of H, with respect to u, is zero. This is found by calculating

$$\lim_{\epsilon \to 0} \frac{1}{\epsilon}(H(u + \epsilon v, \lambda, \gamma) - H(u, \gamma, \lambda))$$

$$= \int_0^T \left(u(t) - \sum_{i=1}^N \lambda_i \ell_{t_i}(t) + \sum_{i=1}^N \gamma_i \ell_{t_i}(t) \right) v(t)dt.$$

Setting this expression equal to zero gives that the optimal $u^\star$ is given by

$$u^\star(t) = \sum_{i=1}^{N} \lambda_i \ell_{t_i}(t) - \sum_{i=1}^{N} \gamma_i \ell_{t_i}(t) = \lambda^T \ell(t) - \gamma^T \ell(t), \qquad (3.18)$$

where, as before,

$$\ell(t)^T = (\ell_{t_1}(t), \ell_{t_2}(t), \ldots, \ell_{t_N}(t))^T.$$

We now eliminate $u^\star$ from H to obtain

$$H(u^\star, \lambda, \gamma)$$

$$= \frac{1}{2} \int_0^T ((\lambda^T - \gamma^T)\ell(t))^2 dt + \sum_{i=1}^{N} \lambda_i(a_i - L_{t_i}(u^\star)) - \sum_{i=1}^{N} \gamma_i(L_{t_i}(u^\star) - b_i)$$

$$= \frac{1}{2}(\lambda - \gamma)^T G(\lambda - \gamma) + \lambda^T a - \gamma^T b - \sum_{i=1}^{N} \lambda_i L_{t_i}(u^\star) + \sum_{i=1}^{N} \gamma_i L_{t_i}(u^\star)$$

$$= \frac{1}{2}(\lambda - \gamma)^T G(\lambda - \gamma) + \lambda^T a - \gamma^T b - \sum_{i=1}^{N}\sum_{j=1}^{N} \lambda_i \lambda_j L_{t_i}(\ell_{t_j})$$

$$+ \sum_{i=1}^{N}\sum_{j=1}^{N} \lambda_j \gamma_i L_{t_i}(\ell_{t_j}) + \sum_{i=1}^{N}\sum_{j=1}^{N} \lambda_i \gamma_j L_{t_i}(\ell_{t_j}) - \sum_{i=1}^{N}\sum_{j=1}^{N} \gamma_j \gamma_i L_{t_i}(\ell_{t_j})$$

$$= \frac{1}{2}(\lambda - \gamma)^T G(\lambda - \gamma) + \lambda^T a - \gamma^T b - \lambda^T G\lambda + \lambda^T G\gamma + \lambda^T G\gamma - \gamma^T G\gamma$$

$$= \frac{1}{2}(\lambda - \gamma)^T G(\lambda - \gamma) + \lambda^T a - \gamma^T b - (\lambda - \gamma)^T G(\lambda - \gamma)$$

$$= -\frac{1}{2}(\lambda - \gamma)^T G(\lambda - \gamma) + \lambda^T a - \gamma^T b.$$

We can thus write $H(u^\star, \lambda, \gamma)$ in a form suitable for use in quadratic programming in the following manner:

$$H(u^\star, \lambda, \gamma) \qquad (3.19)$$

$$= -\frac{1}{2}\begin{pmatrix} \lambda^T & \gamma^T \end{pmatrix}\begin{pmatrix} G & -G \\ -G & G \end{pmatrix}\begin{pmatrix} \lambda \\ \gamma \end{pmatrix} + \begin{pmatrix} \lambda^T & \gamma^T \end{pmatrix}\begin{pmatrix} a \\ -b \end{pmatrix}.$$

Thus, to find the optimal $u^\star$, we need only solve the quadratic programming problem

$$\max_{\lambda, \gamma} H(u^\star, \lambda, \gamma), \qquad (3.20)$$

subject to the component-wise positivity constraints on λ and γ.

In general, the control laws constructed by this technique are going to drive the control close to the endpoints of the interval. A better control law can be obtained by penalizing the control for deviating from the center of the interval. The derivation is essentially the same in concept, but it is a bit more complex in calculation.

3.4 SMOOTHING SPLINES

In many problems, to insist that the control drive the output through the points of the data set is overly restrictive, and in fact leads to control laws that produce wild excursions of the output between the data points. This phenomenon was observed by Wahba, who developed a theory of smoothing splines that corrected this problem to a certain degree.

In this section, we will develop a theory of smoothing splines based on the same optimal control techniques used to produce interpolating splines. However, we will penalize the control for missing the data points instead of imposing hard constraints.

3.4.1 Problem 3

Problem 3: Smoothing Splines

Let

$$J(u) = \sum_{i=1}^{N} w_i (L_{t_i}(u) - \alpha_i)^2 + \rho \int_0^T u(t)^2 dt.$$

The problem is

$$\min_{u \in L_2} J(u).$$

The constants w_i, $i = 1, \ldots, N$, are assumed to be strictly positive, as is the smoothing parameter ρ. The choice of the parameters w_i and ρ is important. They control the rate of convergence of the optimal control signal as the number of data points goes to infinity. This is discussed in detail in [91]. The choice of the smoothing parameter ρ for fixed data sets is an important issue and is discussed at length in the monograph by Wahba [96].

Even though Problem 3 can be solved quite elegantly as a minimum norm problem in Hilbert spaces, the topic of the next chapter, we here follow the variational method used in the previous section in order to stress that this important problem admits multiple solution methods.

We first calculate the Gateaux derivative of J in the form

$$
\begin{aligned}
\lim_{\epsilon \to 0} \frac{1}{\epsilon}(J(u + \epsilon v) - J(u)) & \\
&= \sum_{i=1}^{N} 2w_i L_{t_i}(v)(L_{t_i}(u) + \alpha_i) + 2\rho \int_0^T v(t)u(t)dt \\
&= 2 \int_0^T \left[\sum_{i=1}^{N} w_i \ell_{t_i}(t)(L_{t_i}(u) + \alpha_i) + \rho u(t) \right] v(t)dt. \quad (3.21)
\end{aligned}
$$

Now, to ensure that u is a minimum, we must have that the Gateaux derivative vanishes along all directions v, but this can only happen if

$$
\sum_{i=1}^{N} w_i \ell_{t_i}(t)(L_{t_i}(u) + \alpha_i) + \rho u(t) = 0. \quad (3.22)
$$

To simplify matters, we consider the operator $\mathcal{T} : L_2[0, T] \to L_2[0, T]$, given by

$$
[\mathcal{T}(u)](t) = \sum_{i=1}^{N} w_i \ell_{t_i}(t) L_{t_i}(u) + \rho u(t), \quad (3.23)
$$

or, in an equivalent form,

$$
[\mathcal{T}(u)](t) = \int_0^T \left(\sum_{i=1}^{N} w_i \ell_{t_i}(t) \ell_{t_i}(s) \right) u(s)ds + \rho u(t). \quad (3.24)
$$

Our goal is to show that the operator $\mathcal{T}$ is one-to-one and onto.

Lemma 3.3 *The operator $\mathcal{T}$ is one-to-one for all choices of $w_i > 0$, $i = 1, \ldots, N$, and $\rho > 0$.*

Proof. Suppose $\mathcal{T}(u_0) = 0$ for some $u_0 \in L_2$. (3.23) directly gives that

$$
\sum_{i=1}^{N} w_i \ell_{t_i}(t) L_{t_i}(u_0) + \rho u_0(t) = 0,
$$

and hence that

$$
\sum_{i=1}^{N} w_i \ell_{t_i}(t) \xi_i + \rho u_0(t) = 0,
$$

where ξ_i is the constant $L_{t_i}(u0)$. This implies that any solution u_0 of $T(u_0) = 0$ is in the span of the set $\{\ell_{t_i}, \ i = 1, \ldots, N\}$.

Hence, consider a solution of the form

$$u_0(t) = \sum_{i=1}^{N} \tau_i \ell_{t_i}(t)$$

and evaluate $T(u_0)$ to obtain

$$\sum_{i=1}^{N} w_i \ell_{t_i}(t) L_{t_i} \left(\sum_{j=1}^{N} \tau_j \ell_{t_j}(t) \right) + \rho \sum_{i=1}^{N} \tau_i \ell_{t_i}(t) = 0.$$

Thus, for each i,

$$w_i \sum_{j=1}^{N} L_{t_i}(\ell_{t_j}) \tau_j + \rho \tau_i = 0.$$

The coefficient τ $(\tau^T = (\tau_1, \ldots, \tau_N)^T)$ is then the solution of a set of linear equations of the form

$$(WG + \rho I)\tau = 0,$$

where W is the diagonal matrix of the weights w_i and $G = [g_{ij}]$ is, as before, the Grammian with $g_{ij} = L_{t_i}(\ell_{t_j})$.

Now, consider the matrix $WG + \rho I$ and multiply it on the left by W^{-1}, and consider the scalar

$$z^T(G + \rho W^{-1})z = z^T G z + \rho z^T W^{-1} z > 0,$$

since both G and ρW^{-1} are positive definite. Thus, for positive weights and positive ρ, the only solution is $\tau = 0$. ∎

It remains to show that the operator T is onto.

Lemma 3.4 *For $\rho > 0$ and $w_i > 0$, $i = 1, \ldots, N$, the operator T is onto.*

Proof. Suppose T is not onto. Then there exists a nonzero function $\omega \in L_2[0, T]$ such that

$$\int_0^T \omega(t)[T(u)](t)dt = 0,$$

for all $u \in L_2[0, T]$. We have, after some manipulation,

$$\int_0^T \omega(t)[T(u)](t)dt$$

$$= \int_0^T \left[\int_0^T \sum_{i=1}^{N} w_i \ell_{t_i}(t) \ell_{t_i}(s) \omega(t)dt + \rho \omega(s) \right] u(s)ds = 0,$$

and hence

$$\int_0^T \sum_{i=1}^N w_i \ell_{t_i}(t) \ell_{t_i}(s) \omega(t) dt + \rho \omega(s) = 0, \ \forall s \in [0, T].$$

By the previous lemma, the only solution of this equation is $\omega = 0$, and hence $\mathcal{T}$ is onto. ∎

We have arrived at the following proposition.

Proposition 3.5 *The functional*

$$J(u) = \sum_{i=1}^N w_i (L_{t_i}(u) + \alpha_i)^2 + \rho \int_0^T u^2(t) dt$$

has a unique, global minimum.

Proof. We use (3.22) to find the optimal solution to Problem 3. As in the proof that $\mathcal{T}$ is one-to-one, we look for a solution of the form

$$u(t) = \sum_{i=1}^N \tau_i \ell_{t_i}(t).$$

Substituting this expression into (3.22), we have upon equating coefficients of $\ell_{t_i}(t)$, $i = 1, \dots, N$, the system of linear equations $(WG + \rho I)\tau = W\alpha$. As in the proof of Lemma 3.3, the coefficient matrix is invertible, and hence the solution exists and is unique. ∎

The resulting curve $y(t)$ is a spline. The major difference between classic, interpolating splines and smoothing splines is that the nodal points are determined by the optimization instead of being predetermined. It should, moreover, be noted that inverting the matrix $WG + \rho I$ is not trivial. Since it is a Grammian, we can expect it to be badly conditioned. However, by using the techniques in [30], the conditioning can be improved.

As an example, consider again the example in Subsection 3.2.2, but with the cost functional in Problem 3 with $\rho = 5 \cdot 10^{-6}$, $w_1 = \cdots = w_4 = 1$. The resulting smoothing spline is shown in Figure 1.2.

3.5 SMOOTHING SPLINES WITH CONSTRAINTS

In this section, we consider two different problems. The first problem that we will consider is a rather straightforward extension of Problem 2. The derivation, though, is significantly more involved. This is stated as Problem 4, and the resulting spline is of significant practical importance.

The second class of problems we consider in this chapter, namely, problems involving stochastic data, is very important. It is often the case that there is something known about the underlying curve; for example, in SD we may have some prior knowledge about the function f. For example, if the data represent growth data on a child from age three months to seven years, we can be reasonably assured that the function f is monotonously increasing, and the resulting spline must also be monotonously increasing if the curve is to have any credibility. There are also cases in which the underlying curve is convex, or more information is known so that the curve must satisfy other shape constraints. In this section, we will show that the techniques we have developed for optimal control can be used to formulate and solve a version of these problems.

3.5.1 Problem 4

Problem 4: Smoothing Splines with Constraints
Let the cost functional be defined by

$$J(u) = \frac{1}{2} \int_0^T u^2(t)dt + \frac{1}{2} \sum_{i=1}^N (L_{t_i}(u) - \zeta_i)^2,$$

where

$$\zeta_i = \frac{a_i + b_i}{2}.$$

The problem is

$$\min_{u \in L_2} \ J(u),$$

subject to the constraints of Problem 2.

An example of a curve that is obtained through Problem 4 is given in Figure 3.1.

To solve Problem 4, define H as

$$H(u, \lambda, \gamma) = \frac{1}{2} \int_0^T u^2(t)dt + \frac{1}{2} \sum_{i=1}^N (L_{t_i}(u) - \zeta_i)^2$$
$$+ \sum_{i=1}^N \lambda_i(a_i - L_{t_i}(u)) + \sum_{i=1}^N \gamma_i(L_{t_i}(u) - b_i). \quad (3.25)$$

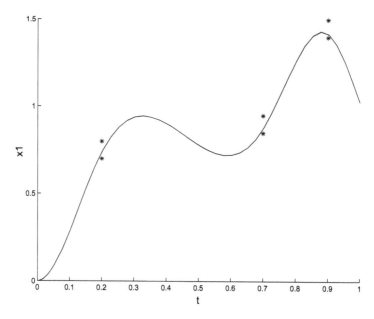

Figure 3.1 Interpolating through intervals while penalizing deviations from the midpoints of the intervals. Here, a second-order system with both eigenvalues equal to -1 was used to produce the scalar output $y = x_1$, with $\dot{x}_1 = -x_1 + x_2$, $\dot{x}_2 = -x_2 + u$.

As before, we want to minimize H with respect to u and maximize with respect to λ and γ. Calculating the Gateaux derivative of H with respect to u, we find

$$\lim_{\epsilon \to 0} \frac{1}{\epsilon} (H(u + \epsilon v, \lambda, \gamma) - H(u, \lambda, \gamma)) \tag{3.26}$$
$$= \int_0^T \left[u(t) + \sum_{i=1}^N \ell_{t_i}(t)(L_{t_i}(u) - \zeta_i) - \sum_{i=1}^N \lambda_i \ell_{t_i}(t) + \sum_{i=1}^N \gamma_i \ell_{t_i}(t) \right] v(t) dt.$$

Setting this equal to 0, we find the condition that

$$u(t) + \sum_{i=1}^N \ell_{t_i}(t)(L_{t_i}(u) - \zeta_i) - \sum_{i=1}^N \lambda_i \ell_{t_i}(t) + \sum_{i=1}^N \gamma_i \ell_{t_i}(t) = 0. \tag{3.27}$$

Thus, we see, once again, that we must have the optimal u as a linear combination of the ℓ_{t_i},

$$u^\star(t) = \sum_{i=1}^N \tau_i \ell_{t_i}(t). \tag{3.28}$$

This, again, has the effect of reducing the nonparametric problem to a problem of calculation of parameters in a finite-dimensional space.

Substituting $u^\star$ into H, we have

$$
\begin{aligned}
H(\tau, \lambda, \gamma) &\qquad\qquad (3.29)\\
= \frac{1}{2}\tau^T G\tau + \frac{1}{2}\tau^T G^2\tau - \tau^T G(a+b) + \lambda^T a - \tau^T G\lambda\\
+ \tau^T G\gamma - \gamma^T b + \kappa,
\end{aligned}
$$

where κ is a constant that does not affect the location of the optimal point. The problem is now the following:

$$
\max_{\lambda\geq 0,\gamma\geq 0}\ \min_{\tau\in\mathbb{R}^N}\ H(\zeta,\lambda,\gamma). \qquad (3.30)
$$

Calculating the derivative of H with respect to τ, we have

$$
\frac{\partial H}{\partial \tau} = G\tau^\star + G^2\tau^\star - G(a+b) - G\lambda + G\gamma = 0, \qquad (3.31)
$$

where $\tau^\star$ is optimal. Solving for $\tau^\star$, we have

$$
\tau^\star = (I+G)^{-1}(a+b+\lambda-\gamma). \qquad (3.32)
$$

Now, substituting this into H, we get

$$
\begin{aligned}
H(\tau^\star, \lambda, \gamma) &\qquad\qquad (3.33)\\
= \frac{1}{2}(a+b+\lambda-\gamma)^T G(I+G)^{-2}(a+b+\lambda-\gamma)\\
+ \frac{1}{2}(a+b+\lambda-\gamma)^T G^2(I+G)^{-2}(a+b+\lambda-\gamma)\\
- (a+b+\lambda-\gamma)^T (I+G)^{-1}G(a+b+\lambda-\gamma)\\
+ \lambda^T a - \gamma^T b + \kappa\\
= -\frac{1}{2}\left(\begin{array}{cc}\lambda^T & \gamma^T\end{array}\right)\left(\begin{array}{cc}(I+G)^{-1}G & -(I+G)^{-1}G\\ -(I+G)^{-1}G & (I+G)^{-1}G\end{array}\right)\left(\begin{array}{c}\lambda\\ \gamma\end{array}\right)\\
+ \left(\begin{array}{cc}\lambda^T & \gamma^T\end{array}\right)\left(\begin{array}{c}(I+G)^{-1}G(a+b)+a\\ -(I+G)^{-1}G(a+b)-b\end{array}\right) + \kappa_2,
\end{aligned}
$$

where κ_2 is a constant.

What remains is to solve the quadratic programming problem

$$
\begin{aligned}
\max_{\lambda\geq 0,\gamma\geq 0}\Bigg\{ &-\frac{1}{2}\left(\begin{array}{cc}\lambda^T & \gamma^T\end{array}\right)\left(\begin{array}{cc}(I+G)^{-1}G & -(I+G)^{-1}G\\ -(I+G)^{-1}G & (I+G)^{-1}G\end{array}\right)\left(\begin{array}{c}\lambda\\ \gamma\end{array}\right)\\
&+ \left(\begin{array}{cc}\lambda^T & \gamma^T\end{array}\right)\left(\begin{array}{c}(I+G)^{-1}G(a+b)+a\\ -(I+G)^{-1}G(a+b)-b\end{array}\right)\Bigg\}.\qquad (3.34)
\end{aligned}
$$

This problem is easily solved numerically and produces splines which are quite well behaved. Although the quadratic programming problem is more complicated in terms of the matrices, these formulations seem to offer enough improvement over the formulation of Problem 2 to be worthwhile.

3.5.2 Problem 5

The next problem we consider is the problem of constructing monotone splines. This problem, as we discussed in the beginning of this section, is very important for many practical applications. We will do less than construct monotone splines here, although it is possible to extend the techniques we are using to produce an infinite-dimensional quadratic programming problem that produces monotone splines. We do not make that extension in this section, but restrict ourselves to ensuring that the spline is nondecreasing at each node. This problem has a significant increase in difficulty over the problems we have considered to this point.

Problem 5: Monotone Smoothing Splines
Let

$$J(u) = \rho \int_0^T u^2(t)dt + \sum_{i=1}^N w_i(L_{t_i}(u) - \alpha_i)^2,$$

and let a set of constraints be imposed as

$$DL_{t_i}(u) \geq 0, \ i = 1, \ldots, N.$$

The problem then becomes

$$\min_{u \in L_2} \ J(u).$$

We define H as

$$H(u, \lambda) = J(u) + \frac{1}{2}\sum_{i=1}^N w_i(L_{t_i}(u) - \alpha_i)^2 - \sum_{i=1}^N \lambda_i DL_{t_i}(u). \quad (3.35)$$

As before, the idea is to minimize H over u and to maximize H over all component-wise positive Lagrange multipliers $\lambda \in \mathbb{R}^N$. The scheme is to construct the control that minimizes H as a function of λ, and to use this parameterized control to convert H to a function of a finite set of parameters. The resulting H will then be minimized with respect to a subset of the parameters and H will be reduced to a function of λ alone. Then the problem reduces to a quadratic programming problem that can be solved using standard software.

We will use the notation

$$h_{t_i}(s) = \frac{d}{dt}\ell_{t_i}(s)$$

for

$$\frac{d}{dt}\ell_t(s)\Big|_{t=t_i},$$

and we can then rewrite H as

$$H(u,\lambda) = \frac{1}{2}\int_0^T \left(u^2(s) - \sum_{i=1}^N \lambda_i \frac{d}{dt}\ell_{t_i}(s)u(s) \right) ds + \sum_{i=1}^N w_i(L_{t_i}(u) - \alpha_i)^2.$$

$$(3.36)$$

We calculate the Gateaux derivative of H with respect to u in the direction $v \in L_2$ to obtain

$$D_u H(u,\lambda)(v) \hspace{4cm} (3.37)$$

$$= \int_0^T \left(u(s) - \sum_{i=1}^N \lambda_i h_{t_i}(s) + \sum_{i=1}^N w_i(L_{t_i}(u) - \alpha_i)\ell_{t_i}(s) \right) v(s) ds,$$

and thus the optimal u must satisfy

$$u(s) - \sum_{i=1}^N \lambda_i h_{t_i}(s) + \sum_{i=1}^N w_i(L_{t_i}(u) - \alpha_i)\ell_{t_i}(s) = 0. \hspace{1cm} (3.38)$$

As a consequence, the optimal u can be represented as

$$u(s) = \sum_{i=1}^N \lambda_i h_{t_i}(s) + \sum_{i=1}^N \tau_i \ell_{t_i}(s), \hspace{1cm} (3.39)$$

and the representation is unique provided that, for each i, $i = 1,\ldots,N$, the functions h_{t_i} and ℓ_{t_i} are linearly independent. This linear independence condition reduces to the condition that $A^{n-1} \neq 0$, and with the conditions of (3.3) that $A \neq 0$. So we may assume, without loss of generality, that the representation is unique.

We now substitute u into H to get a function of τ and λ. We first establish some notation which we will need in order to simplify the formulation. Let

$$M = [h_{ij}], \quad h_{ij} = \int_0^T h_{t_i}(s)h_{t_j}(s)ds,$$

$$K = [k_{ij}], \quad k_{ij} = \int_0^T h_{t_i}(s)\ell_{t_j}(s)ds,$$

$$G = [g_{ij}], \quad g_{ij} = \int_0^T \ell_{t_i}(s)\ell_{t_j}(s)ds.$$

We now substitute the expression for u into $H(u, \lambda)$ and obtain, after considerable simplification,

$$H(\tau, \lambda) = \frac{1}{2}\left[\lambda^T M\lambda + \lambda^T K\tau + \tau^T K^T\lambda + \tau^T G\tau\right] - \lambda^T M\lambda - \lambda^T K\tau$$

$$+ \frac{1}{2}\sum_{i=1}^N w_i \left[\lambda^T K^T e_i e_i^T K\lambda + \tau^T G e_i e_i^T G\tau + \alpha_i^2\right.$$

$$+ \quad 2\lambda^T K^T e_i e_i^T G\tau - 2\alpha_i \lambda^T K^T e_i - 2\alpha_i \tau^T G e_i\Big],$$

where e_i is the ith unit vector in $\mathbb{R}^N$. We rewrite this after further simplification as

$$H(\tau, \lambda) = -\frac{1}{2}\lambda^T M\lambda + \frac{1}{2}\tau^T G\tau + \frac{1}{2}\lambda^T K^T W K\lambda + \frac{1}{2}GWG\tau$$

$$+ \frac{1}{2}\alpha^T W\alpha + \lambda^T KWG\tau - \lambda^T K^T W\alpha - \tau^T GW\alpha,$$

where, as before, W is the diagonal matrix of weights w_i.

We now calculate the Gateaux derivative of H with respect to τ and obtain

$$D_\tau H(\tau, \lambda)(z) = z^T(G\tau + GWG\tau + GWK^T\lambda - G\alpha). \tag{3.40}$$

Setting this equal to zero, we have that the optimal τ must satisfy the equation

$$(G + GWG)\tau = G\alpha - GWK\lambda, \tag{3.41}$$

or equivalently, since G is invertible,

$$\tau = (W^{-1} + G)^{-1}(\alpha - K\lambda). \tag{3.42}$$

It is clear that when we substitute this into H we have a quadratic function of λ, so the problem is reduced to solving a quadratic programming problem. Some simplification is possible, but the overall form of the matrices involved becomes rather messy. In fact, we get the general expression for $H(\lambda)$ as

$$H(\lambda) = \lambda^T F_1 \lambda + F_2 \lambda + F_3 \alpha. \tag{3.43}$$

It is easy enough to generalize this construction to include higher-order derivative constraints. The only problem that arises is to ensure that the representation of the optimal control in (3.39) is unique. If we were to generalize the construction to linear combinations of derivatives and to different linear combinations at different points, this obstruction becomes quite severe, and some of the same problems arise here as do in the case of Birkhoff interpolation, as discussed in [106].

Note that this construction does not guarantee that the spline function is monotone, but only that the function is nondecreasing at each node. In numerical experiments, we have found that by adding points we can create monotone splines using this construction. We have not, however, at this point proved that the addition of a finite number of points suffices to produce monotone splines. In Chapter 7 we will return to this important problem.

3.6 DYNAMIC TIME WARPING

Traditionally, statistics has dealt with discrete data sets. However, most statisticians would agree that information is sometimes lost when data are considered to be point or vector data. In longitudinal studies, it is clear that it is the record of an event that is important, not the individual measurements. For example, if one is studying the growth of individuals in an isolated community, it is not the heights at yearly intervals but the curve that represents the growth of an individual over a sequence of years that is of interest. These matters are discussed at length in the seminal book of Ramsay and Silverman [80]. This book makes a very convincing argument for the study of curves as opposed to discrete data sets.

Often when studying curves it is not clear that the independent variable (which we will refer to as time) is well defined. Li and Ramsay [58] consider several examples which make this point quite well. This problem has also arisen when trying to construct weight curves for premature babies– the time of conception is seldom known exactly, and different ethnic groups may have different growth curves. This leads to the problem of "dynamic time warping" or "curve registration" in order to compare curves that have different bases of time. In this section, we will follow the development of Li and Ramsay [58] and of Ramsay and Silverman [80]. We will show that their formulation is equivalent to the problem of optimal output tracking in control theory, and then give a formulation that is somewhat better behaved from an optimization point-of-view.

3.6.1 Time-like Functions

Consider a set of curves

$$DC = \{f_i(t) \in \mathcal{F} \mid i = 1, \ldots, N\},$$

and assume that there is some commonality among the curves encoded through the set $\mathcal{F}$, for example, they are all growth curves. Choose one such curve, say $f_0(t)$. The choice of this curve is discussed in some detail in [80].

Let

$$\mathcal{T} = \{\xi_\delta(t) \mid \xi_\delta(t) \text{ is "time-like"}\},$$

parameterized by δ in some compact set D. What we mean by "time-like" is quite vague. We would like for the functions to be at least almost monotone, although at this point we do not want to impose too many conditions. We can pose the problem in the following manner:

$$\min_{\delta \in D} \|f_0(t) - f_i(\xi_\delta(t))\|.$$

This problem, although elegant in its simplicity, is too general to solve. In [58], the set of pseudo-times is constructed in a very clever and insightful manner. In particular, the condition that

$$\frac{\ddot{\xi}(t)}{\dot{\xi}(t)} = v(t),$$

with $v(t)$ being "small," is imposed. The idea behind this construction is that this will make the curvature of the pseudo-time $\xi(t)$ small and so the resulting function $\xi(t)$ would be time-like.

3.6.2 Problem 6

We can now reformulate the curve registration and dynamic time warping problem in the context of optimal control. (We emphasize that we are only reformulating the problem in [58].)

Problem 6: Dynamic Time Warping

$$\min_{u \in L_2} \int_0^T (u^2(t) + (f_i(t) - f_0(x(t)))^2) dt,$$

subject to the constraint

$$\ddot{x}(t) = u(t)\dot{x}(t).$$

The question of initial data is problem dependent, and we let it fall outside the scope of this discussion. However, the problem can be solved with or without the initial data being given. (See Polak [77] for a complete treatment of this issue.) And, rather than actually solving Problem 6, we simply note that this is an important problem one would typically like to be able to solve.

3.6.3 Problem 7

Now, consider a control system with output, of the form

$$\begin{aligned}
\dot{x}(t) &= f(x(t)) + u(t)g(x(t)), \\
y(t) &= h(x(t)),
\end{aligned} \tag{3.44}$$

and let a curve $z(t)$ be given which we would like to follow as closely as possible with the output curve $y(t)$. The classical model following problem is to construct u so that the distance between y and z is minimized. This problem is often considered in the asymptotic sense, but in reality the finite time domain is the most important in almost all applications.

Problem 7: Model Following

$$\min_{u \in L_2} \int_0^T (u^2(t) + (z(t) - y(t))^2)dt,$$

subject to the constraint that

$$\begin{aligned}
\dot{x}(t) &= f(x(t)) + u(t)g(x(t)), \\
y(t) &= h(x(t)).
\end{aligned}$$

It is clear that Problem 6 is a special case of Problem 7 by taking

$$\tilde{z} = \begin{pmatrix} x \\ \dot{x} \end{pmatrix},$$

and then noting that

$$\dot{\tilde{z}} = \begin{pmatrix} 0 & 1 \\ 0 & 0 \end{pmatrix} \tilde{z} + u \begin{pmatrix} 0 & 0 \\ 0 & 1 \end{pmatrix} \tilde{z},$$

with $z = (1\,0)\tilde{z}$.

The solution to this problem is given by the solution to the corresponding Euler-Lagrange equations, which in this case is a nonlinear two-point boundary value problem. The nonlinearity comes from the fact that the differential equation has a nonlinearity; in Problem 6 the equation is bilinear and in Problem 7 the equation is nonlinear affine. However, even if we choose a linear constraint, the Euler-Lagrange equations are nonlinear because the output function is a nonlinear function of the state. These problems are in general only solvable by numerical methods. (Again, Polak [77] is useful here.)

3.7 TRAJECTORY PLANNING

The trajectory planning problem is a fundamental problem in aeronautics, robotics, biomechanics, and many other areas of application in control theory. The problem comes in two distinct versions. The most general version is of interest for autonomous vehicles or autonomous movement in general. There the complete route is not known in advance and must be planned "online." This problem is far beyond what we can do with the relatively simple tools we have developed here. The version of the problem we will discuss in this section is in contrast quite simple. We are given a sequence of target points and target times, and we are required to be close to the point at some time close to the target time.

The problem of being close in space is nicely solved by Problems 2 and 4, but the problem of being close in time has been difficult to resolve. The concept of "dynamic time warping" seems to be the tool that can resolve it. Neither the problem nor its solution is trivial, and it is unfortunate that there does not appear to be an analytic solution.

3.7.1 Problem 8

We define an auxiliary system which we will use as the pseudo-time. We have chosen the system to be linear rather than the more complicated nonlinear system of Li and Ramsay. Let

$$F = \begin{pmatrix} 0 & 1 \\ 0 & 1 \end{pmatrix},$$

$g^T = (0, 1)$, and $h = (1, 0)$. We then consider the system

$$\begin{aligned} \dot{z} &= Fz + gv, \\ \xi &= hz, \end{aligned} \qquad (3.45)$$

with 0 initial data. We then have the output ξ represented as

$$\xi(t) = t + \int_0^T (t - s)v(s)ds,$$

so that if v is small, ξ is indeed "time-like."

We now formulate the trajectory planning problem in the following manner:

Problem 8: Trajectory Planning

Let

$$J(u, v) = \frac{1}{2} \sum_{i=1}^{N} w_i (L_{\xi(t_i)}(u) - \alpha_i)^2$$

$$+ \frac{1}{2} \int_0^T (u^2(t) + \rho v^2(t)) dt + \frac{1}{2} \sum_{i=1}^{N} w_i (\xi(t_i) - t_i)^2.$$

The problem is

$$\min_{v \in L_2} \min_{u \in L_2} J(u, v),$$

subject to the constraints

$$\dot{z} = Fz + gv,$$
$$\xi = hz,$$

and

$$\dot{\xi}(t_i) \geq 0.$$

There is no really clean solution to this problem, but we are able to present an algorithm which gives at least a suboptimal solution. We first minimize with respect to u (this is just Problem 3), and we find that the optimal u is of the form

$$u(t) = \sum_{i=1}^{N} \tau_i \ell_{\xi(t_i)}(t), \tag{3.46}$$

where the τ_i are chosen as in the solution to Problem 3. Recall that the vector τ satisfies the matrix equation

$$(DW + \rho I)\tau = W\alpha,$$

and hence the τ_i are functions of the as of yet nonoptimal $\xi(t_i)$. However, this choice of u is optimal for any choice of $\xi(t_i)$. Substituting u into the functional J, we reduce J to a function of v alone. We are then faced with a highly nonlinear functional to be minimized subject to a differential equation constraint. That is, we now have an optimal control problem with a nonlinear cost functional and a very simple linear control system as the constraint. It is possible to write the Euler-Lagrange equations for this problem, but they are somewhat intimidating. Instead, we opt for a suboptimal solution that has a chance of being calculated using good numerical optimization procedures.

We assume that the control v will have the form

$$v(t) = \sum_{i=1}^{N} \gamma_i (t_i - t)_+, \qquad (3.47)$$

where the function $(t_i - t)_+$ is the standard function of polynomial splines that is zero when $t > t_i$ and is $t_i - t$ otherwise, and the γ_i are to be determined. Calculating $\xi(t)$ with this choice of v, we see that ξ is a cubic spline with nodes at times t_i. Substituting ξ into the cost functional, we have that J is now a function of the N parameters γ_i, $i = 1, \ldots, N$. In other words, we have reduced the problem to the a finite-dimensional optimization problem. At this point, we have not taken into account the inequality constraints on the derivatives of the $\xi(t_i)$; this is done by introducing the Hamiltonian, H, in Problem 5. Because of the nonquadratic nature of the cost functional, the quadratic programming problem has become a nonlinear programming problem, and, as such, presents more difficult numerical problems.

SUMMARY

In this chapter, eight different problems originating from optimal control theory, statistics, and numerical analysis were introduced. We showed that these problems can be formulated and, to a certain degree, addressed within a unified framework based on the relationship between optimal control and conventional or smoothing splines.

The first five problems concerned the minimization of a quadratic cost functional subject to a set of linear constraints, ranging from exact interpolation or penalized deviations from the nodes, to splines that pass through intervals or are nondecreasing at the nodes. For these problems, we were able to come up with explicit solutions.

In Problems 6 and 7, the curve registration problem was addressed by extending our optimal control formulation to include the concept of "dynamic time warping." The last problem concerned trajectory planning where, given a set of target points and target times, we wanted to be close to the points at times close to the target times. We were able to formulate these problems within our optimal control framework in order to unite them with the previous problems into one unified theory.

In the coming chapters, we will return to a number of these problems and see how they can be viewed directly as minimum norm problems in certain Hilbert spaces, as well as discuss extensions and potential applications. Thus, this chapter should be thought of as hinting of things to come, rather than presenting a self-contained set of solutions.

Chapter Four

SMOOTHING SPLINES AND GENERALIZATIONS

Estimation and smoothing for data sets that contain random data present difficulties not present in deterministic data sets. Yet such data sets are very common in practice and, if the nature of the data is not respected, conclusions may be drawn that have little relation to reality. This book presents a treatment that unifies the concepts of interpolating and smoothing splines, and, in this chapter we will study the basic construction–the smoothing spline–as a minimum norm problem in a suitable Hilbert space. This approach unifies a series of problems addressed in [31],[33],[69],[91],[101], [106]. Furthermore, the approach of this chapter gives a unified treatment of smoothing splines as developed by Wahba [96] and the classical polynomial and exponential interpolating splines. The approach of this chapter rests on the Hilbert space methods developed by Luenberger in [64].

The theory of smoothing splines is based on the premise that a datum, α, is the sum of a deterministic part, β, and a random part, ϵ. It is assumed that ϵ is the value of a random variable drawn from some probability distribution. Smoothing splines are designed to approximate the deterministic part by minimizing the variance of the random part. Often the random variable comes from measurement error. We start this chapter with two examples in which the random error comes either from the measurements or from estimations based on incomplete data.

Example 4.1 *A seemingly straightforward problem is to determine the volume of water contained in a playa lake in West Texas [89]. These are transient water supplies that, because of their formation, are almost perfectly circular. If a transect is made across the center of the lake, it is possible to obtain a fairly good estimate of the volume. At the boundary of the lake the depth of the water is 0 cm. However, the depth is typically measured by a person (e.g., a graduate student) wading through the lake and measuring the depth at a series of points. The bottom of the lake is silted and so it is not clear where the bottom of the probe rests. The measurements are made by* *reading the depth of a marked probe. These measurements are indeed quite random. The data set then consists of two deterministic values at the boundary, and a series of random numbers representing the depth at a series of predetermined points.*

Example 4.2 *For most individuals in the United States, their home is the principal component of their financial portfolio. The question of the value of the portfolio is of interest in a variety of economic indicators [70]. When the home is purchased, there is a firm monetary value that can be measured and when the home is sold, there is another firm value. In between, the value is less certain. Almost every individual can give you an estimate of the value, but unless a formal appraisal is done there may be a very large error in the estimate. This results in a data set with a few deterministic values, the purchase price, the selling price, and formal appraisals, and many random values that are estimates by the owner.*

These two examples have in common some data that can be assumed to be exact and some data that are subject to error. In this chapter, we will consider the problem of approximating discrete data using the dynamics of a linear controlled system. The system may have hard constraints such as boundary values or hard constraints at internal values. The data will be assumed to be noisy with known statistics. A contribution of this chapter is to formulate these problems as a general class of minimum norm problems in Hilbert spaces, as formulated in [69],[91],[105]. The advantage of this formulation is more than conceptual in that the smoothed data are immediately available, as is the smooth functional approximation. Thus we are able to split the problems into an estimation problem and a problem of constructing interpolating splines on a derived data set. Both of these problems permit fast numerical solutions.

As before, we will let

$$\dot{x} = Ax + bu, \quad y = cx, \quad x(0) = x_0 \qquad (4.1)$$

be a controllable and observable linear system, with initial data $x(0) = x_0$. We will think of this system as the curve generator. As will be seen, we achieve the smoothest approximation if we impose the conditions for $n \geq 2$

$$cb = cAb = cA^2b = \cdots = cA^{n-2}b = 0, \qquad (4.2)$$

where n is the dimension of the system.

Now, let the data set be given as

$$D = \{(t_i, \alpha_i) : i = 1, \ldots, N\},$$

and assume that $t_i > 0$ and let $T = t_N$. We will refer to the points t_i as the "nodes." Our goal is to find a control $u(t)$ that minimizes

$$J(u, x_0) = \int_0^T u^2(t)dt + (\hat{y} - \hat{\alpha})^T Q(\hat{y} - \hat{\alpha}) + x_0^T R x_0, \qquad (4.3)$$

where Q and R are positive definite matrices. It is not strictly necessary for these matrices to be positive definite. However, as in the case of the linear regulator, if they are not positive definite then other conditions must be imposed to ensure a unique solution. We will discuss this further in Section 4.1.1. The vector $\hat{y} \in \mathbb{R}^N$ has components

$$y_i = y(t_i) = ce^{At_i}x_0 + \int_0^{t_i} ce^{A(t_i-s)}bu(s)ds,$$

and, similarly, the vector $\hat{\alpha} \in \mathbb{R}^N$ has components α_i.

As before, we find it convenient to define the basis functions

$$\ell_i(s) = \begin{cases} ce^{A(t_i-s)}b, & t_i \geq s, \\ 0, & t_i < s. \end{cases} \qquad (4.4)$$

Note that if the assumption in (4.2) holds, then $\ell_i(s)$ is $n-2$ times continuously differentiable at t_i, that is,

$$\ell_i^{(k)}(t) = \begin{cases} cA^k e^{A(t_i-s)}b, & t_i \geq s, \\ 0, & t_i < s. \end{cases} \qquad (4.5)$$

As long as $cA^k b = 0$, $\ell_i^{(k)}(t)$ is continuous, which by (4.2) holds until $k = n-2$. We can now write

$$y_i = ce^{At_i}x_0 + \int_0^T \ell_i(s)u(s)ds = ce^{At_i}x_0 + \langle \ell_i, u \rangle_{L_2},$$

where $\langle \ell_i, u \rangle_{L_2} := \int_0^T \ell_i(s)u(s)ds$.

Now, let $\beta_i := R^{-1}e^{A^T t_i}c^T$. Then

$$y_i = ce^{At_i}x_0 + \int_0^T \ell_i(s)u(s)ds = \langle \beta_i, x_0 \rangle_R + \langle \ell_i, u \rangle_{L_2},$$

where we define the inner product $\langle x, w \rangle_R = x^T R w$.

Note that if we take the derivatives of $y(t)$, when $u = \ell_i(t)$, we have

$$y^{(2n-2)}(t) = \sum_{k=0}^{n-2} cA^{n-2+k}b\ell_i^{(n-2-k)}(t) + \int_0^T cA^{2n-1}e^{A(t-s)}b\ell_i(s)ds.$$

This derivative is continuous, but the next derivative fails to be so. Thus, this particular y is $2n-2$ times continuously differentiable everywhere, and real analytic between the nodes.

4.1 THE BASIC SMOOTHING PROBLEM

In this section we state the basic problem of smoothing splines and construct the solution. Here we show that the construction splits into two parts
in a very natural way. Ultimately, this will allow the implementation of fast
algorithms for smoothing spline constructions. The basic idea of the construction is to define a linear variety, in a Hilbert space, that is defined by the
constraints. The data are then defined as a point in the Hilbert space, and the
optimization reduces to finding the point on the affine variety that is closest
(in the sense of the norm in the Hilbert space) to the data point. We know
that we can construct this point by finding the orthogonal complement of the
linear variety that defines the affine variety and constructing the intersection
of the affine variety with the orthogonal complement. In this process we
follow Luenberger [64].

4.1.1 The Hilbert Space and the Affine Variety

Let

$$\mathcal{H} = L_2[0, T] \times \mathbb{R}^n \times \mathbb{R}^N,$$

with norm

$$\|(u; x; d)\|_{\mathcal{H}}^2 = \int_0^T u^2(t)dt + d^T Q d + x^T R x$$

and corresponding inner product

$$\langle (u; x; d), (v; z; f) \rangle_{\mathcal{H}} = \int_0^T u(t)v(t)dt + x^T R z + d^T Q f.$$

Now, since the output at time t_i is

$$y_i = \langle \beta_i, x_0 \rangle_R + \langle \ell_i, u \rangle_{L_2},$$

we can define the linear subspace of constraints in $\mathcal{H}$, V_0, encoding the
dynamics, as

$$V_0 = \{(u; x; d) : \quad 0 = -d_i + \langle \beta_i, x \rangle_R + \langle \ell_i, u \rangle_{L_2}, \ i = 1, \dots, N\}.$$

We use the notation V_0 for consistency with later notation. (Note that V_0
is of infinite dimension since it contains a copy of $L_2[0, T]$, and is of finite
codimension since it is the intersection of a finite number of codimension 1
subspaces.)

With this notation, the basic smoothing spline problem becomes

$$\min_{(u;x;d) \in \mathcal{H}} \|(u; x; d) - (0; 0; \hat{\alpha})\|_{\mathcal{H}}^2, \tag{4.6}$$

such that $(u; d; x) \in V_0$. We will use p to denote the point $(0; 0; \hat{\alpha}) \in \mathcal{H}$.
We will now construct the orthogonal complement of V_0 in $\mathcal{H}$.

Lemma 4.3 *The orthogonal complement of V_0 in $\mathcal{H}$ is*

$$V_0^\perp = \left\{ (v; w; z) \ \middle| \ w + \sum_{i=1}^N \langle z, e_i \rangle_Q \beta_i = 0, \ \ v + \sum_{i=1}^N \langle z, e_i \rangle_Q \ell_i = 0 \right\},$$

where e_i is the ith unit vector in $\mathbb{R}^N$.

Proof. By definition,

$$V^\perp = \{ (v; w; z) \mid \langle v, u \rangle_{L_2} + \langle z, d \rangle_Q + \langle w, x \rangle_R = 0, \forall \, (u; x; d) \in V_0 \}.$$

We have

$$\langle z, d \rangle_Q = \sum_{i=1}^N \langle z, e_i \rangle_Q d_i = \sum_{i=1}^N \langle z, e_i \rangle_Q [\langle \beta_i, x \rangle_R + \langle \ell_i, u \rangle_{L_2}]$$

$$= \left\langle \sum_{i=1}^N \langle z, e_i \rangle_Q \beta_i, x \right\rangle_R + \left\langle \sum_{i=1}^N \langle z, e_i \rangle_Q \ell_i, u \right\rangle_{L_2}.$$

Therefore,

$$0 = \langle v, u \rangle_{L_2} + \langle w, x \rangle_R + \langle z, d \rangle_Q$$

$$= \langle v, u \rangle_{L_2} + \langle w, x \rangle_R + \left\langle \sum_{i=1}^N \langle z, e_i \rangle_Q \beta_i, x \right\rangle_R + \left\langle \sum_{i=1}^N \langle z, e_i \rangle_Q \ell_i, u \right\rangle_{L_2}$$

$$= \left\langle w + \sum_{i=1}^N \langle z, e_i \rangle_Q \beta_i, x \right\rangle_R + \left\langle v + \sum_{i=1}^N \langle z, e_i \rangle_Q \ell_i, u \right\rangle_{L_2}.$$

From the definition of V_0 we have that, given a pair $(u : x_0)$, there exists a
d so that $(u; x_0; d) \in V_0$. Thus, the above equality is true for $u \in L_2[0, T]$
and for all $x \in \mathbb{R}^n$ and, as a consequence, we must have that

$$0 = w + \sum_{i=1}^N \langle z, e_i \rangle_Q \beta_i \quad \text{and} \quad 0 = v + \sum_{i=1}^N \langle z, e_i \rangle_Q \ell_i.$$

Note that the latter equality is in the sense of $L_2[0, T]$. The lemma thus
follows. ∎

4.1.2 The Intersection

Before constructing the intersection $V_0 \cap (V_0^\perp + p)$, two things must be verified, namely, that V_0 is nonempty and closed. That V_0 is nonempty is a consequence of the fact that every choice of u and x determines a triple in V_0. We state as a lemma the fact that V_0 is closed.

Lemma 4.4 V_0 *is a closed subspace of the Hilbert space* $\mathcal{H}$.

Proof. Define the function with domain $L_2[0, T] \times \mathbb{R}^n$ and range $\mathbb{R}^N$ as

$$K_i((u; x)) = \langle \beta_i, x \rangle_R + \langle \ell_i, u \rangle_{L_2}. \tag{4.7}$$

Note that K_i is continuous since it is defined in terms of the inner products, and note that V_0 is the graph of K, where K is the function with components K_i. It then follows from the closed graph theorem that V_0 is closed in $\mathcal{H}$. ∎

Since V_0 is closed, we have that the intersection of V_0 and $V_0^\perp + p$ consists of a single point. This point is the solution to the optimal control problem in (4.3).

Lemma 4.5 *The intersection of* $V_0 \cap (V_0^\perp + p)$ *is*

$$V_0 \cap (V_0^\perp + p) = \left\{ \left(\sum_{i=1}^N \gamma_i \ell_i; \sum_{i=1}^N \rho_i \beta_i; (I + GQ + FQ)^{-1}(GQ + FQ)\hat{a} \right) \right\},$$

where G *is the Grammian of the* β_i^T, F *is the Grammian of the* ℓ_i, *and where*

$$\gamma_i = \langle [I - (I + GQ + FQ)^{-1}(GQ + FQ)]\hat{a}, e_i \rangle_Q,$$
$$\rho_i = \langle [I - (I + GQ + FQ)^{-1}(GQ + FQ)]\hat{a}, e_i \rangle_Q.$$

Proof. Equating quantities from V_0 and $V_0^\perp + p$ (here $p = (0; 0; \hat{a})$ is the data point), we have from the definition of V_0 and some rearrangement of terms

$$d_i = \langle \beta_i, x \rangle_R + \langle \ell_i, u \rangle_{L_2}$$
$$= -\sum_{j=1}^N \langle z, e_j \rangle_Q \langle \beta_i, \beta_j \rangle_R - \sum_{j=1}^N \langle z, e_j \rangle_Q \langle \ell_i, \ell_j \rangle_{L_2}.$$

Now, equating d with $\hat{y}$ and z with $\hat{y} - \hat{a}$, we get

$$y_i = -\sum_{j=1}^N \langle \hat{y} - \hat{a}, e_j \rangle_Q \langle \beta_i, \beta_j \rangle_R - \sum_{j=1}^N \langle \hat{y} - \hat{a}, e_j \rangle_Q \langle \ell_i, \ell_j \rangle_{L_2}$$
$$= -e_i^T GQ(\hat{y} - \hat{a}) - e_i^T FQ(\hat{y} - \hat{a}).$$

Note that since the ℓ_i are linearly independent, F is invertible.

In more compact form, we have

$$\hat{y} = -(GQ + FQ)(\hat{y} - \hat{\alpha}),$$

or, finally, we have that

$$(I + GQ + FQ)\hat{y} = (GQ + FQ)\hat{\alpha}. \tag{4.8}$$

By rewriting $I + GQ + FQ = (Q^{-1} + F + G)Q$, and since F and Q are positive definite and G is positive semidefinite, the matrix $(I + GQ + FQ)$ is invertible and we find $\hat{y}$ as a linear function of the data $\hat{\alpha}$ as

$$\hat{y} = (I + GQ + FQ)^{-1}(GQ + FQ)\hat{\alpha}.$$

This $\hat{y}$ is the optimal smoothed estimate of the data $\hat{\alpha}$. Using $\hat{y}$ we can then calculate both the optimal control and the optimal initial condition using the defining equations of the orthogonal complement.

To construct the optimal control u^*, we have from Lemma 4.3 and the identifications above:

$$u^*(t) = -\sum_{i=1}^{N} \langle \hat{y} - \hat{\alpha}, e_i \rangle_Q \ell_i(t)$$

$$= -\sum_{i=1}^{N} \langle (I + GQ + FQ)^{-1}(GQ + FQ)\hat{\alpha} - \hat{\alpha}, e_i \rangle_Q \ell_i(t)$$

$$= \sum_{i=1}^{N} \langle [I - (I + GQ + FQ)^{-1}(GQ + FQ)b]\hat{\alpha}, e_i \rangle_Q \ell_i(t).$$

The construction of the optimal initial condition is carried out in a similar manner. Thus the lemma is proved. ∎

We summarize the results of the section with the following theorem.

Theorem 4.6 *Let*

$$\dot{x} = Ax + bu, \quad y = cx$$

be a controllable and observable linear system with initial data $x(0) = x_0$, let a data set be given as

$$D = \{(t_i, \alpha_i) \mid i = 1, \ldots, N\},$$

and assume that $t_i > 0$ and let $T = t_N$. Let the cost function be given as

$$J(u, x_0) = \int_0^T u^2(t)dt + (\hat{y} - \hat{\alpha})^T Q(\hat{y} - \hat{\alpha}) + x_0^T R x_0,$$

where Q and R are positive definite matrices. The vector $\hat{y}$ has components

$$y_i = y(t_i) = c e^{At_i} x_0 + \int_0^{t_i} c e^{A(t_i-s)} b u(s) ds,$$

and the vector $\hat{\alpha}$ has components α_i. Minimizing J over $u \in L_2[0, t]$ and $x_0 \in \mathbb{R}^n$, the optimal smoothed data are given by

$$\hat{y} = (I + GQ + FQ)^{-1}(GQ + FQ)\hat{\alpha}, \qquad (4.9)$$

the optimal control is given by

$$u = \sum_{i=1}^{N} \langle [I - (I + GQ + FQ)^{-1}(GQ + FQ)]\hat{\alpha}, e_i \rangle_Q \ell_i, \qquad (4.10)$$

and the optimal initial condition is given by

$$x_0 = \sum_{i=1}^{N} \langle [I - (I + GQ + FQ)^{-1}(GQ + FQ)]\hat{\alpha}, e_i \rangle_Q \beta_i. \qquad (4.11)$$

4.2 THE BASIC ALGORITHM

In the previous section, we formulated and solved a problem using an algorithm that is, in some sense, just an implementation of the Hilbert projection theorem. This algorithm is extremely powerful. We will see in this book many problems in optimal control that can be solved using this algorithm. We will now formally state the algorithm with some explanation, as a parallel to the discussion in Chapter 2. We begin by describing the inputs and outputs of the algorithm.

INPUTS:

- a quadratic cost function in the control, possibly the initial data, and the data;

- a given set of constraints that includes a linear control system, and deterministic constraints on the solution of the control system, and the initial data.

OUTPUTS:

- smoothed data $\hat{y}$;

- optimal control u;

- optimal initial data x_0.

THE ALGORITHM

1. Define the Hilbert space of the control, initial data, and data as $\mathcal{H} = \mathcal{H}_1 \times \mathcal{H}_2 \times \mathcal{H}_3$, where $\mathcal{H}_1$ is the Hilbert space of the control, $\mathcal{H}_2$ is the Hilbert space of the initial data, and $\mathcal{H}_3$ is the Hilbert space of the data, which may be finite- or infinite-dimensional. The norm is based on the cost functional.

2. Define the affine subvariety V_c of the constraints. In many of the applications c is replaced by the parameter from the problem.

3. Define the data through the point p in $\mathcal{H}$.

4. Verify that the variety is well defined in the Hilbert space. Verify that any point evaluations are well defined.[1]

5. Verify that the variety is closed. This is an essential step but usually in these problem is a direct consequence of the closed graph theorem.

6. Calculate the orthogonal complement of V_0. This step may or may not be complicated. It is usually straight forward.

7. Calculate the intersection of $(V_0^\perp + p) \cap V_c$. This step can be complicated because it reduces to solving a system of linear equations derived from the definitions of V_c, $V_0^\perp$, and $V_0^\perp + p$. The equations can be a mix of integral equations and finite dimensional linear equations, and may involve several parameters that must be eliminated.

8. The solution to the equations exist and is unique since we know that the intersection will contain a single point. This point is the optimal u, x_0 and the optimal, smoothed output of the linear system.

To a certain degree, this book can be thought of as an exercise in applying this algorithm to solve a series of important problems in the theory of control theoretic smoothing splines. There are of course other methods for solving these problems. However, no other method seems as straightforward and intuitive. It is basically just a generalization of the problem from Euclidean geometry of finding a point on a given line nearest to a given point in the plane–a problem from high school geometry. (See [64] for many other such examples.)

---

[1] Since $L_2[0, T]$ is a Hilbert space of equivalence classes, point evaluations, for example, $f(\alpha)$, are not well defined. However, the inner product $\langle \ell_t, f \rangle = \int_0^T \ell_t(s) f(s) ds$ is well defined as a function of t.

In the next section we will explore the application of this algorithm to a series of problems about splines in various forms. In this chapter we will assume that the Hilbert space of data is finite-dimensional, and will consider the more general case in Chapter 6.

4.3 INTERPOLATING SPLINES WITH INITIAL DATA

For interpolating splines we are required to find a control that drives the output y through the points in the data set D. This can be expressed in terms of additional constraints of the form

$$\alpha_i = \langle \beta_i, x_0 \rangle_R + \langle \ell_i, u \rangle_{L_2}$$

for $i = 1, \ldots, N$. The goal is to find a control and an initial condition that minimize

$$J(u, x_0) = \int_0^T u^2(t)dt + x_0^T R x_0$$

subject to the constraints. Just as for smoothing splines, we let the Hilbert space be

$$\mathcal{H} = L_2[0, T] \times \mathbb{R}^n.$$

Now, the affine variety of constraints is given by

$$V_{\hat{\alpha}} = \{(u; x_0) : 0 = -\alpha_i + \langle \beta_i, x_0 \rangle_R + \langle \ell_i, u \rangle_{L_2}, \ i = 1, \cdots, N\}.$$

Here the goal is to find the point in $V_{\hat{\alpha}}$ of minimum norm. The procedure is much the same as for smoothing splines. We first must verify that $V_{\hat{\alpha}}$ is nonempty. This follows from the hypothesized controllability of the linear system. We construct $V_0^{\perp}$ and construct the intersection

$$V_0^{\perp} \cap V_{\hat{\alpha}},$$

which consists of a single point (see [64]), provided that V_0 is closed.

Lemma 4.7 V_0 *is closed.*

Proof. Let $K_i(x, w) = \langle \beta_i, x \rangle_R + \langle \ell_i, w \rangle_{L_2}$. Now K_i is a continuous linear functional on the Hilbert space $\mathcal{H}$ and hence $\ker(K_i)$ is a closed subset of $\mathcal{H}$. Moreover, V_0 is the intersection of a finite number of closed subsets and hence is closed. ∎

After some calculation we have

$$V_0^\perp = \left\{ (v; w) \;\middle|\; v = \sum_{i=1}^{N} \tau_i \ell_i, \, w = \sum_{i=1}^{N} \tau_i \beta_i \right\},$$

which gives that the optimal u is in fact given by

$$u = \sum_{i=1}^{N} e_i^T (F + G)^{-1} \hat{\alpha} \ell_i,$$

and the optimal initial condition is given by

$$x_0 = \sum_{i=1}^{N} e_i^T (F + G)^{-1} \hat{\alpha} \beta_i.$$

Here, the matrices F and G, the vectors β_i, and the elements of $L_2[0, T]$, $\ell_i(t)$, are as in the previous section. This is just a slight generalization of the construction given in [91], and hence the details are left out.

For cubic splines, the classical construction reduces to solving a system of equations of the form $Mx = \rho$, where M is tridiagonal. In [100], the construction of interpolating splines is reduced to solving banded matrices. However, in both cases, additional constraints are required to make the problem have a unique solution. With the procedure developed here, the additional constraints are unnecessary because of the optimization. Neither classical cubic splines nor the procedure developed in [100] can easily handle the optimal initial data.

4.4 PROBLEMS WITH ADDITIONAL CONSTRAINTS

In a series of papers, Willsky and coauthors [2],[3], and Krener [57] considered an estimation problem based on a stochastic boundary value problem. In this section, we consider a similar problem in which the smoothing spline is generated by a linear system for which there are hard constraints. The constraints may occur as boundary values, but they may also occur as fixed internal values, or even as linear operator constraints on the solution. We will show that many of these problems can be formulated and solved with the machinery we have established. The basic idea is that we have a data set in which each data point is of the form $\alpha_i = f(t_i) + \epsilon_i$, where $f(t_i)$ is deterministic and ϵ_i is the value of a random variable. The goal is to produce a curve (the spline) that better approximates $f(t)$. This is, of course, a standard statistical formulation (see [96]).

4.4.1 Two-Point Boundary Value Problems

We begin by considering a general boundary value problem. Let the boundary condition be given by

$$\Phi x(0) + \Psi x(T) = h, \tag{4.12}$$

where we let $h \in \mathbb{R}^k$. This, of course, includes the classical two-point boundary value formulations and other problems of interest. We note that, since

$$x(T) = e^{AT} x(0) + \int_0^T e^{A(T-s)} bu(s)ds,$$

the specific dependence on $x(T)$ can be removed and the boundary constraint simply becomes

$$Px(0) + \Psi \int_0^T e^{A(T-s)} bu(s)ds = h, \tag{4.13}$$

where

$$P = \Phi + \Psi e^{AT}.$$

Note that if there is any solution to (4.12), then by the controllability hypothesis there is a solution to (4.13). We hypothesize that there is at least one solution of (4.12).

Define the Hilbert space as

$$\mathcal{H} = L_2[0, T] \times \mathbb{R}^n \times \mathbb{R}^N,$$

with norm

$$\|(u; x_0; y)\|_{\mathcal{H}}^2 = \int_0^T u^2(t)dt + x_0^T R x_0 + y^T Q y,$$

and define the constraint variety to be

$$V_h = \Big\{ (u; x; d) \ \Big| \ d_i = \langle \beta_i, x \rangle_R + \langle \ell_i, u \rangle_{L_2},$$
$$Px + \Psi \int_0^T e^{A(T-s)} bu(s)ds = h \Big\}.$$

We first prove the following lemma.

Lemma 4.8 V_0 *is a closed subspace of* $\mathcal{H}$.

Proof. The mapping

$$(u; x) \rightarrow \Psi \int_0^T e^{A(T-s)} bu(s) ds + Px$$

with domain $L_2[0, T] \times \mathbb{R}^n$ is continuous, and hence the subspace

$$W = \left\{ (u, x) \in L_2[0, T] \times \mathbb{R}^n \,\middle|\, Px + \Psi \int_0^T e^{A(T-s)} bu(s) ds = 0 \right\}$$

is closed. Now the mapping from W to $\mathbb{R}^N$ defined by

$$d_i = \langle \beta_i, x \rangle_R + \langle \ell_i, u \rangle_{L_2},$$

which also is continuous, and again we appeal to the closed graph theorem to finish the proof. ∎

We now construct $V_0^\perp$.

Lemma 4.9 *For some* $\lambda \in \mathbb{R}^k$,

$$V_0^\perp = \left\{ (v; w; z) \,\middle|\, w = -\sum_{i=1}^N \langle z, e_i \rangle_Q \beta_i + R^{-1} P^T \lambda, \right.$$
$$\left. v = -\sum_{i=1}^N \langle z, e_i \rangle_Q \ell_i + (\Psi e^{A(T-t)})^T \lambda \right\}.$$

Proof. The first part of the construction is exactly the same as in subsection 4.1.1, and from there we have

$$V_0^\perp = \left\{ (v; w; z) \,\middle|\, \langle w + \sum_{i=1}^N \langle z, e_i \rangle_Q \beta_i, x \rangle + \langle v + \sum_{i=1}^N \langle z, e_i \rangle_Q \ell_i, u \rangle = 0 \right\}.$$

This relationship does not hold for all x and u, but only for those x and u for which (4.13) holds. Multiplying by λ^T, $\lambda \in \mathbb{R}^k$, we can rewrite (4.13) as

$$\langle R^{-1} P^T \lambda, x \rangle_R + \langle (\Psi e^{A(T-t)})^T \lambda, u \rangle_{L_2} = 0. \tag{4.14}$$

From this we conclude that

$$w + \sum_{i=1}^N \langle z, e_i \rangle_Q \beta_i = R^{-1} P^T \lambda, \quad v + \sum_{i=1}^N \langle z, e_i \rangle_Q \ell_i = (\Psi e^{A(T-t)})^T \lambda,$$

and the lemma follows. ∎

It remains to construct the intersection $V_h \cap (V_0^\perp + p)$ to find the optimal solution. This construction is technically more complicated than the basic smoothing spline but the underlying technique is identical.

The unique point in the intersection is defined as the solution of the following system of four equations in the unknowns u, x_0, y, and λ, obtained by identifying x and w with x_0, d with $\hat{y}$, and z with $\hat{y} + \hat{\alpha}$.

$$u = -\sum_{i=1}^{N} \langle \hat{y} - \hat{\alpha}, e_i \rangle_Q \ell_i + b^T e^{A^T(T-t)} \Psi^T \lambda, \tag{4.15}$$

$$x_0 = -\sum_{i=1}^{N} \langle \hat{y} - \hat{\alpha}, e_i \rangle_Q \beta_i + R^{-1}(\Phi + \Psi e^{A^T})^T \lambda, \tag{4.16}$$

$$h = P x_0 + \int_0^T \Psi e^{A(T-s)} b u(s) ds, \tag{4.17}$$

$$y_i = \langle \beta_i, x_0 \rangle_R + \langle \ell_i, u \rangle_{L_2}. \tag{4.18}$$

We begin by eliminating x_0 and u from (4.18) by substituting (4.15) and (4.16). After some manipulation we have

$$y_i = e_i^T G(\hat{y} - \hat{\alpha}) - e_i^T F(\hat{y} - \hat{\alpha}) + \beta_i^T P^T \lambda + \int_0^T \ell_i(s) b^T e^{A^T(T-s)} \Psi^T ds \lambda.$$

Since $\beta_i = R^{-1} e^{A^T t_i} c^T$, let

$$\beta = R^{-1} (e^{A^T t_1} c^T, \dots, e^{A^T t_N} c^T) =: R^{-1} E$$

to obtain

$$\hat{y} = -G(\hat{y} - \hat{\alpha}) - F(\hat{y} - \hat{\alpha}) + E^T R^{-1} P^T \lambda + \Lambda \lambda,$$

where

$$\Lambda = \int_0^T l(s) b^T e^{A^T(T-s)} \Psi^T ds.$$

We will now use (4.17) to obtain a second equation in λ and $\hat{y}$.

$$h = P\left[-\sum_{i=1}^{N} \langle \hat{y} - \hat{\alpha}, e_i \rangle_Q \beta_i + R^{-1} P^T \lambda \right]$$

$$+ \int_0^T \Psi e^{A(T-s)} b \left[-\sum_{i=1}^{N} \langle \hat{y} - \hat{\alpha}, e_i \rangle_Q \ell_i + b^T e^{A^T(T-s)} \Psi^T \lambda \right] ds.$$

We make the following observation:

$$\sum_{i=1}^{N} \langle \hat{y} - \hat{\alpha}, e_i \rangle_Q \beta_i = \sum_{i=1}^{N} \beta_i e_i^T Q(\hat{y} - \hat{\alpha}) = R^{-1} E Q(\hat{y} - \hat{\alpha}).$$

We now define

$$M = \sum_{i=1}^{N} \int_0^T \Psi e^{A(T-s)} b \ell_i(s) e_i^T \, ds Q,$$

and hence

$$\sum_{i=1}^{N} \int_0^T \Psi e^{A(T-s)} b \langle \hat{y} - \hat{a}, e_i \rangle_Q \ell_i(s) \, ds = M(\hat{y} - \hat{a}).$$

Using these two constructions, we then have

$$h = P(-R^{-1}EQ(\hat{y} - \hat{a})) + PR^{-1}P^T \lambda - M(\hat{y} - \hat{a}) + \Psi \Gamma \Psi^T \lambda, \quad (4.19)$$

where Γ is the controllability Grammian

$$\Gamma = \int_0^T e^{A(T-s)} b b^T e^{A^T(T-s)} \, ds.$$

By combining these two expressions, linking $\hat{y}$ and λ gives the following system of linear equation

$$\begin{pmatrix} I + (G+F)Q & -E^T R^{-1} P^T - \Lambda \\ PR^{-1}EQ - M & PR^{-1}P^T + \Psi \Gamma \Psi^T \end{pmatrix} \begin{pmatrix} \hat{y} \\ \lambda \end{pmatrix}$$
$$= \begin{pmatrix} (G+F)Q\hat{a} \\ h + PR^{-1}EQ + M\hat{a} \end{pmatrix}. \quad (4.20)$$

Note first that the matrix on the left is invertible because of the uniqueness and existence of the solution to the minimum norm problem. Using (4.20) we can solve for $\hat{y}$ and for λ. These values can be used in (4.15) and (4.16) to uniquely determine the optimal control and optimal initial condition. As before, we see that the optimal estimate of the data is obtained independently of the control.

Remark: The matrix E is a Grammian-like matrix that determines whether the initial data can be recovered from sampled observational data, that is, if $\dot{x} = Ax$, $x(0) = \zeta$, $y = cx$, and the output is sampled at a set of discrete points t_i, then the output is recoverable from these observations if and only if E has full rank. Thus E plays the same role as the observability Grammian. There are no known necessary and sufficient conditions for E to have full rank. This problem was studied originally by Smith and Martin and was reported in [88]. It is also interesting that the controllability Grammian arises in the formulation of (4.19). The reason for the controllability Grammian to appear is more obvious when one considers the simpler problem of optimally moving between affine subspaces. This problem is studied in Chapter 8.

4.4.2 Multiple Point Constraints

In this case we have a hard constraint of the form

$$\Phi_1 x(r_1) + \cdots + \Phi_k x(r_k) = h,$$

and the data set

$$D = \{(t_i, \alpha_i) \mid i = 1, \ldots, N\},$$

and we assume, without loss of generality, that

$$\{r_i \mid i = 1, \ldots, k\} \cap \{t_i \mid i = 1, \ldots, N\} = \emptyset.$$

We again make the assumption that there exists at least one set of vectors a_i such that

$$\Phi_1 a_1 + \cdots + \Phi_k a_k = h.$$

We construct the variety of constraints and note that we can replace $x(r_i)$ with

$$e^{Ar_i} x(0) + \int_0^{r_i} e^{A(r_i - s)} b u(s) ds.$$

Thus the constraint depends only on u and x_0. We use the Hilbert space

$$\mathcal{H} = L_2[0, T] \times \mathbb{R}^n \times \mathbb{R}^N,$$

and the constraint variety V_h is

$$\left\{ (u; x_0; \hat{y}) \,\middle|\, y_i = \langle \beta_i, x_0 \rangle + \langle \ell_i, u \rangle_{L_2}, \right.$$
$$\left. \sum_{i=1}^{k} \Phi_i e^{Ar_i} x_0 + \sum_{i=1}^{k} \int_0^T \Phi_i \ell_{r_i}(s) u(s) ds = h \right\}.$$

As before, we construct the orthogonal complement to V_0 and then determine the intersection

$$V_h \cap (V_0^\perp + (0; 0; \hat{a})).$$

We leave this construction to the reader.

4.4.3 Examples

In this section we will present some examples of problems that fit this generalized boundary value formulation. We let

$$A = \begin{pmatrix} 0 & 1 \\ 0 & 0 \end{pmatrix}, \ b = \begin{pmatrix} 0 \\ 1 \end{pmatrix}, \ c = (\ 1 \ \ 0 \ ), \ T = 1, \quad (4.21)$$

$$t_1 = 0.2, \ t_2 = 0.3, \ t_3 = 0.5, \ t_4 = 0.7, \ t_5 = 0.8, \quad (4.22)$$

$$\hat{\alpha} = (\ 0.8 \quad 0.2 \quad 0.5 \quad 1 \quad 0.3 \ ), \quad (4.23)$$

$$Q = 10^4 I_5, \ R = 10^4 I_2, \ (I_p = p \times p \ \text{identity matrix}). \quad (4.24)$$

Example 4.10 (*periodic splines*) *We first study the situation when we insist that $x(0) = x(T)$. In this case we have that $\Phi = -\Psi = I_2$, while $h = 0$. The solution is depicted in Figure 4.1.*

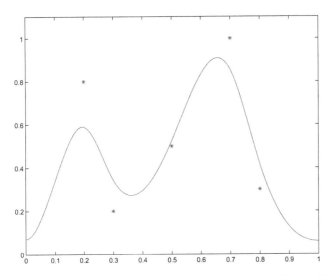

Figure 4.1 Periodic splines. Here the boundary value is given by $x(0) = x(T)$. Depicted are $y(t)$ (solid) and α_i, $i = 1, \ldots, 4$ (stars).

Example 4.11 (*two-point boundary value problems*) *We now let the boundary constraint be encoded by $\Phi = (1, 1)$, $\Psi = -\Phi$, $h = 1$, which implies that the boundary values are given by the set*

$$\{(x_0, x_T) \mid (1, 1)x_0 - (1, 1)x_T = 1\}.$$

The resulting output curve that solves this problem is shown in Figure 4.2.

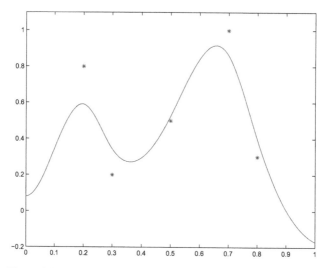

Figure 4.2 Boundary value problem: $(1,1)x(0) - (1,1)x(T) = 1$.

4.4.4 Integral Constraints

In many applications ranging from statistics to medicine there are constraints of the form

$$\int_0^1 y(t)\, dt = 1.$$

We will consider a simple problem with $\dot{x} = Ax + bu$ and a data set $D = \{(t_i, \alpha_i) \mid i = 1, \ldots, N\}$, and we will further assume that each $\alpha_i > 0$. Our constraint variety, V_1, is given by

$$\Big\{(u; x_0; y) \, \Big| \, y_i = \langle \beta_i, x_0 \rangle_R + \langle \ell_i, u \rangle_{L_2},$$
$$1 = \int_0^T y(t)dt, \ y(t) = ce^{At}x_0 + \int_0^t e^{A(t-s)}bu(s)ds\Big\}.$$

As per the algorithm in Section 4.2, we compute $V_0^\perp$. The definition of the orthogonal complement gives

$$V_0^\perp = \{(v; w; z) \mid \langle v, u \rangle + \langle_{L_2} w, x_0 \rangle_R + \langle z, y \rangle_q = 0, \forall (u; x_0, y) \in V_0\}.$$

Using the defining relation (after some calculation using the first relationship in the definition of V_0) we have

$$\left\langle v + \sum_{i=1}^N \langle z, e_i \rangle_Q \ell_i, u \right\rangle_{L_2} + \left\langle w + \sum_{i=1}^N \langle z, e_i \rangle_Q \beta_i, x_0 \right\rangle_R = 0. \quad (4.25)$$

From the second and third defining relations for V_0 we have

$$0 = \int_0^T \left(e^{At}x_0 + \int_0^t e^{A(t-s)}bu(s)ds \right)dt$$

$$= \int_0^T e^{At}dt x_0 + \int_0^T \int_s^T e^{A(t-s)}bdt u(s)ds,$$

and multiplying both sides by λ gives

$$0 = \left\langle \int_0^T R^{-1}e^{A^T t}dt\lambda, x_0 \right\rangle_R + \left\langle \int_s^T b^T e^{A(t-s)}dt\lambda, u \right\rangle_{L_2}. \qquad (4.26)$$

Now, using (4.25) and $V_0^\perp$, we have

$$V_0^\perp = \left\{ (v; w; z) \mid v = -\sum_{i=1}^N \langle z, e_i \rangle_Q \ell_i + \int_s^T b^T e^{A(t-s)}dt\lambda, \right.$$
$$\left. w = -\sum_{i=1}^N \langle z, e_i \rangle_Q \beta_i + \int_0^T R^{-1}e^{A^T t}dt\lambda \right\}.$$

Let $p = (0; 0; \hat{\alpha})$, where $\hat{\alpha}$ is the vector of data. Then, in order to construct $(V_0^\perp + p) \cap V_1$, we must solve the following four equations:

$$y_i = \langle \beta_i, x_0 \rangle_R + \langle \ell_i, u \rangle_{L_2}, \qquad (4.27)$$

$$1 = \int_0^T e^{At}dt x_0 + \int_0^T \int_s^T e^{A(t-s)}bdt u(s)ds, \qquad (4.28)$$

$$u = -\sum_{i=1}^N \langle y + \alpha, e_i \rangle_Q \ell_i + \int_s^T b^T e^{A(t-s)}dt\lambda, \qquad (4.29)$$

$$x_0 = -\sum_{i=1}^N \langle y + \alpha, e_i \rangle_Q \beta_i + \int_0^T R^{-1}e^{A^T t}dt\lambda. \qquad (4.30)$$

The procedure for solving these four equations is exactly the same as before, and we leave the details to the reader. Use (4.29) and (4.30) to eliminate x_0 and u from (4.27) and (4.28). This results in a pair of equations for λ and the optimal y. Solve this system and substitute these values to obtain the optimal u and x_0. After some calculations, the problem of finding the optimal y and λ reduces to solving a matrix equation. The entries in the matrix must be calculated separately and involve some integration that can be done using standard quadrature algorithms. Once y and λ are found, u and x_0 are found by substituting into (4.29) and (4.30.)

SUMMARY

In this chapter we established a common framework for interpolating and smoothing splines via a Hilbert space approach to control theoretic splines.

We demonstrated that control theoretic splines can be used to solve a wide variety of problems. Willsky and colleagues [2],[3] and Krener [57] developed beautiful machinery based on very sophisticated stochastic analysis to solve estimation problems based on stochastic two-point boundary value problems; we showed that the same problems have elegant and simple solutions based on control theoretic splines.

Chapter Five

APPROXIMATIONS AND LIMITING CONCEPTS

One of the desired properties for any smoothing strategy is that it (some-how) converges to the "correct" underlying curve as the number of data points grows. In this chapter we will show that smoothing splines do in fact converge in a certain statistical sense. For this, we will make a few needed assumptions, and we continue to use the earlier assumption that the linear system is controllable and observable as well as the assumption about rel-ative degree. We add a new assumption about the eigenvalues of the state matrix and about the target function, in order to be able to capture the ap-propriate limiting concepts.

5.1 BASIC ASSUMPTIONS

We start by stating the basic assumptions needed to arrive at the main con-vergence result of this chapter:

Assumption 5.1 $c^T b = \cdots = c^T A^{n-2} b = 0$.

Assumption 5.2 *The matrix A has only real eigenvalues.*

Assumption 5.3 *Let the underlying, true curve, $f(t)$, be a C^∞ function on an interval that contains $[0, T]$.*

Assumption 5.4 $x(0) = 0$.

We now suppose that there are infinitely many data points available ob-tained by a repeated sampling of $f(t)$ on the interval $[0, T]$. Let

$$D_N = \{(t_{iN}, \alpha_{iN}) : i = 1, \cdots, N\}$$

be the Nth data set, and let the union of the set of times be dense in the interval $[0, T]$.

We now set

$$y(t) = L_t(u) = \int_0^t c e^{A(t-s)} b u(s) ds,$$

and let u_N be the control that optimizes the functional

$$J_N(u) = \frac{1}{2N} \sum_{i=1}^{N} w_{iN} (L_{t_{iN}}(u) - f(t_{iN}))^2 + \frac{\rho}{2} \int_0^T u^2(t) dt. \qquad (5.1)$$

Here we assume that the weights in the summation are predetermined by the Nth data set.

The control, u_N, was shown to exist uniquely in Chapter 3, and we now want to be able to relate u_N to the control u^* that minimizes

$$J(u) = \frac{1}{2} \int_0^T (L_t(u) - f(t))^2 dt + \frac{\rho}{2} \int_0^T u^2(t) dt. \qquad (5.2)$$

We make the following important assumption:

Assumption 5.5 *The sequence of quadratures defined by the numbers*

$$w_{iN}, t_{iN}$$

converges for all continuous functions defined on $[0, T]$, that is,

$$\lim_{N \to \infty} \frac{1}{2N} \sum_{i=1}^{N} w_{iN} h(t_{iN}) = \frac{1}{2} \int_0^T h(t) dt.$$

Before we are ready to state the main results, a few words should be said about all these assumptions: The assumption about the relative degree is completely reasonable if one is interested in the approximation of curves known only at a few points. In our setting, this makes the output of the system correspond to the position, the second state variable to the velocity, and so on. Thus the control has the effect of controlling the position. This is not necessarily a reasonable assumption if one is interested in trajectory planning, and all of what follows in this chapter can be done in its absence, at the expense of counting derivatives in each theorem. The resulting spline functions will not have as high a degree of differentiability if the assumption is relaxed.

The assumption of real eigenvalues makes certain calculations easier and, as shown in [100], if the eigenvalues are complex, the resulting splines are not nearly as well behaved. We would hesitate to apply these techniques directly to a trajectory planning problem without this assumption.

The assumption of infinite differentiability of the target function can be relaxed, but, if we assume that it is C^k instead, then we will have to count derivatives throughout. (There are probably times in which it would be useful to use functions with step discontinuities, but they can be approximated by C^∞ functions.)

The last assumption on quadrature is critical for understanding the behavior of the spline function in the limit as the number of nodes approaches infinity. The spline functions themselves can be constructed without this assumption but as the number of points increase, we would like to know what is happening to the spline. This assumption tells us that the smoothing part of the cost function behaves in an orderly manner. (There are probably other assumptions that would achieve the same result.)

5.2 CONVERGENCE OF THE SMOOTHING SPLINE

We will prove the following theorem.

Theorem 5.6 *Under assumptions 5.1 to 5.5, the sequence of controls*

$$\{u_N(t)\}_{N=1}^{\infty}$$

converges to the function $u^*(t)$ in L_2-norm, and the sequence of smoothing splines $\{L_t(u_N)\}_{N=1}^{\infty}$ likewise converges to $L_t(u^*)$ in L_2-norm.

Proof outline. We begin the proof by showing that u^* exists and is unique. We show this by explicit construction; we have shown previously that the functions u_N exist and are unique.

Then we will argue that the minimizers of the functionals J_N converge to the minimizer of J. We divide the proof into a series of lemmas.

Lemma 5.7 *The function u^* exists and is unique.*

Proof. We first observe that the cost functional J, given by

$$J(u) = \int_0^T (L_t(u) - f(t))^2 dt + \int_0^T u^2(t) dt, \qquad (5.3)$$

can be reduced to a standard linear-quadratic optimization problem by a change of variable. Let

$$w(t) = L_t(u) - f(t).$$

By taking a sequence of derivatives we have

$$w^{(0)}(t) = \int_0^t c^T e^{A(t-s)} bu(s) ds - f(t), \qquad (5.4)$$

$$w^{(1)}(t) = \int_0^t c^T A e^{A(t-s)} bu(s) ds - f^{(1)}(t), \tag{5.5}$$

$$\vdots$$

$$w^{(n-1)}(t) = \int_0^t c^T A^{n-1} e^{A(t-s)} bu(s) ds - f^{(n-1)}(t), \tag{5.6}$$

$$w^{(n)}(t) = c^T A^{n-1} bu(t) + \int_0^t c^T A^n e^{A(t-s)} bu(s) ds - f^{(n)}(t). \tag{5.7}$$

Here we have used our assumption that $c^T A^k b = 0$ for $k \leq n - 2$. Now let

$$p(t) = t^n - \zeta_{n-1} t^{n-1} - \cdots - \zeta_1 t - \zeta_0$$

be such that $p(A) = 0$. Now, using (5.4) - (5.7), and taking the appropriate weighted sum, we have

$$\begin{aligned} w^{(n)} - \zeta_{n-1} w^{(n-1)} - \cdots - \zeta_0 w \\ = c^T A^{n-1} bu(t) - f^{(n)} + \zeta_{n-1} f^{(n-1)} + \cdots + \zeta_0 f. \end{aligned} \tag{5.8}$$

Writing this in state space form gives

$$\frac{d}{dt} \hat{w}(t) = \hat{A}\hat{w}(t) + (c^T A^{n-1} b) e_n u(t) + (-f^{(n)} + \zeta_{n-1} f^{(n-1)} + \cdots + \zeta_0 f) e_n, \tag{5.9}$$

where $\hat{w} = (w^{(0)}, \ldots, w^{(n-1)})^T$. In (5.9), e_n is the nth unit vector and $\hat{A}$ is in companion form and is similar to A. Let

$$F(s) = (f^n(s) - \zeta_{n-1} f^{n-1}(s) - \cdots - \zeta_0 f(s)).$$

Now, (5.9) defines a closed affine subspace in $L_2[0, T]$ and hence there is a unique function $u^* \in L_2[0, T]$ which gives a point of minimal norm in the affine subspace.   ∎

We now characterize the function u^* of the previous lemma and show that it is at least C^∞.

Lemma 5.8 *The optimal spline $L_t(u^*)$ is given by*

$$L_t(u^*) = \begin{pmatrix} e_1^T & 0 \end{pmatrix} \exp\left(\begin{pmatrix} A & -e_n e_n^T \\ -e_1 e_1^T & -A^T \end{pmatrix} t \right) \begin{pmatrix} \hat{w}(0) \\ \lambda(0) \end{pmatrix}$$

$$- \int_0^t \begin{pmatrix} e_1^T & 0 \end{pmatrix} \exp\left(\begin{pmatrix} A & -e_n e_n^T \\ -e_1 e_1^T & -A^T \end{pmatrix} (t - s) \right) \begin{pmatrix} e_n \\ 0 \end{pmatrix} F(s) ds,$$

and the optimal control is given by

$$u^*(t) = e_n^T \lambda(t),$$

where

$$\frac{d}{dt}\begin{pmatrix} \hat{w}(t) \\ \lambda(t) \end{pmatrix} = \begin{pmatrix} A & -e_n e_n^T \\ -e_1 e_1^T & -A^T \end{pmatrix} \begin{pmatrix} \hat{w}(t) \\ \lambda(t) \end{pmatrix} - \begin{pmatrix} e_n \\ 0 \end{pmatrix} F(s),$$

with data

$$\hat{w}(0) = \hat{w}_0, \quad \lambda(T) = 0.$$

Proof. From the previous lemma we know there is a point of minimal norm in the affine subspace. We will explicitly construct that point using a construction similar to a calculation found in [28].

Define the linear affine subspace

$$Af(\hat{w}_0, F(s)) = \left\{ (\hat{w}; u) \; \middle| \; \hat{w} = \int_0^t e^{\hat{A}(t-s)} (c^T A^{n-1} b) e_n u(s) ds \right\}$$
$$+ \left(e^{\hat{A}t} \hat{w}_0 + \int_0^T e^{\hat{A}(t-s)} F(s) e_n ds, 0 \right). \qquad (5.10)$$

The object is to construct the orthogonal complement to $Af(0,0)$. At this point the construction of the complement is found in [28]. We then construct the intersection of the orthogonal complement to $Af(0,0)$ and $Af(\hat{w}_0, F(s))$. The single point in this intersection is found by solving the two-point boundary value problem

$$\frac{d}{dt}\begin{pmatrix} \hat{w}(t) \\ \lambda(t) \end{pmatrix} = \begin{pmatrix} A & -e_n e_n^T \\ -e_1 e_1^T & -A^T \end{pmatrix} \begin{pmatrix} \hat{w}(t) \\ \lambda(t) \end{pmatrix} - \begin{pmatrix} e_n \\ 0 \end{pmatrix} F(s), \quad (5.11)$$

with data

$$\hat{w}(0) = \hat{w}_0, \quad \lambda(T) = 0,$$

and where

$$u(t) = e_n^T \lambda(t).$$

It is necessary to determine if this two-point boundary value problem has solutions. We solve the differential equation, assuming that it is an initial value problem, to obtain.

$$\begin{pmatrix} \hat{w}(t) \\ \lambda(t) \end{pmatrix} = \exp\left(\begin{pmatrix} A & -e_n e_n^T \\ -e_1 e_1^T & -A^T \end{pmatrix} t \right) \begin{pmatrix} \hat{w}(0) \\ \lambda(0) \end{pmatrix}$$
$$- \int_0^t \exp\left(\begin{pmatrix} A & -e_n e_n^T \\ -e_1 e_1^T & -A^T \end{pmatrix} (t-s) \right) \begin{pmatrix} e_n \\ 0 \end{pmatrix} F(s) ds.$$

Setting $t = T$, we have a linear equation for $\lambda(0)$, and this equation has a unique solution if and only if it has a unique solution for $F = 0$. But for $F = 0$ this is the linear two-point boundary value problem associated with the linear quadratic optimal control problem

$$J(u) = \int_0^T \hat{w}(t)^T e_1 e_1^T \hat{w}(t) + u^2(t) dt$$

with linear constraint

$$\frac{d}{dt} \hat{w} = \hat{A}\hat{w} + c^T A^{n-1} b e_n u(t).$$

This problem has a unique solution since the pair $(e_1, \hat{A})$ is observable and the pair $(\hat{A}, e_n)$ is controllable. It then follows from linear quadratic optimal control theory that the two-point boundary value problem has solutions for all values of $\hat{w}(0)$, and these solutions exist on the interval $[0, T]$. (See, for example, [67].) We note that $\lambda(0) = P(0)\hat{w}(0)$, where $P(t)$ is the solution of the associated Riccati equation. Thus the lemma is proven. $\blacksquare$

We now finish the proof of the theorem by proving convergence. From our assumption that f is C^∞, the control u^* is C^∞ as well. We know that the minimizer of $J(u)$ is unique, and we shown that the minimizer of $J_N(u)$ is unique. We also know from the general theory of optimization [64] that the minimizer of a quadratic functional is given by the unique zero of the Gateaux derivative of the functional. Calculating the Gateaux derivatives, we have the following two linear functionals:

$$DJ(u; w) = \int_0^T (L_t(u) - f(t)) L_t(w) dt + \int_0^T u(t) w(t) dt, \qquad (5.12)$$

$$DJ_N(u; w) = \sum_{i=1}^N w_{iN}(L_{t_{iN}}(u) - f(t_{iN})) L_{t_{iN}}(w) + \int_0^T u(t) w(t) dt. \qquad (5.13)$$

It is clear that, for each u and w, $DJ_N(u; w)$ converges to $DJ(u; w)$ provided the quadrature scheme converges for a sufficiently general class of functions.

We now rewrite the Gateaux derivatives in terms of inner products by the simple expediency of interchanging the order of integration.

$$DJ(u; w) = \int_0^T \left(\int_0^T \ell_t(s)(L_t(u) - f(t)) dt + u(s) \right) w(s) ds, \qquad (5.14)$$

$$DJ_N(u;w) = \int_0^T \left(\sum_{i=1}^N w_{iN}\ell_{t_{iN}}(s)(L_{t_{iN}}(u) - f(t_{iN})) + u(s) \right) w(s)ds.$$

$$(5.15)$$

From the previous two equations, we have that the convergence is independent of w, since

$$\sum_{i=1}^N \ell_{t_{iN}}(s)w_{iN}(L_{t_{iN}}(u) - f(t_{iN})) + u(s) \tag{5.16}$$

converges to

$$\int_0^T \ell_t(s)(L_t(u) - f(t))dt + u(s) \tag{5.17}$$

for every $s \in [0, T]$. We are now concerned with the convergence of linear operators rather than linear functionals.

Let

$$B(s)(u) = \int_0^T \ell_t(s)L_t(u)dt + u(s), \tag{5.18}$$

and define $B_N(s)$ as

$$B_N(s)(u) = \sum_{i=1}^N \ell_{t_{iN}}(s)w_{iN}L_{t_{iN}}(u) + u(s). \tag{5.19}$$

Furthermore, let

$$b(s) = \int_0^T \ell_t(s)f(t)dt$$

and

$$b_N(s) = \sum_{i=1}^N \ell_{t_{iN}}w_{iN}f(t_{iN}).$$

Now, it is clear that $b_N(s)$ converges to $b(s)$ pointwise and hence in L_2-norm. Thus, given ϵ, for N sufficiently large we have

$$|B_N(s)(u_N - u^*)| < \epsilon.$$

Now we know that $B_N(s)x = b_N(s)$ has a unique solution and hence that $B_N(s)$ is nonsingular. We can thus conclude that

$$u_N(s) - u^*(s)$$

converges to 0 pointwise, and since both are smooth and defined on a compact interval in L_2-norm, and the theorem follows. ∎

5.3 QUADRATURE SCHEMES

The theorem from the previous section is important because it shows exactly how the continuous spline is dependent on the data. We see from Theorem 5.6 that the spline is the convolution of the function F, with a kernel that is the semigroup of a Hamiltonian system. We also see that, since the control is optimal with respect to the cost function, the resulting feedback controlled system is stable, and hence that perturbations in F are not blown up, but die quite quickly. This, however, is really just a straightforward result from the theory of linear quadratic optimal control.

We obtain as corollaries three important results.

Corollary 5.9 *Let $w_{iN} = \frac{1}{N}$ and let the sequence of t_i be the observed values of a random variable uniformly distributed in the interval $[0, T]$. Then the sequence of smoothing splines $\{L_t(u_N)\}_{N=1}^{\infty}$ converges to $L_t(u^*)$ in L_2-norm.*

Corollary 5.10 *Let $w_{iN} = \frac{1}{N}$ and let $t_{iN} = \frac{iT}{N}$ (Riemann sum). Then the sequence of smoothing splines $\{L_t(u_N)\}_{N=1}^{\infty}$ converges to $L_t(u^*)$ in L_2-norm.*

Corollary 5.11 *Let w_{iN}, t_{iN} be defined by a Gaussian quadrature scheme. Then the sequence of smoothing splines $\{L_t(u_N)\}_{N=1}^{\infty}$ converges to $L_t(u^*)$ in L_2-norm.*

Proof of Corollaries. The corollaries are presented in the order of the rate of convergence. We begin with Corollary 5.9. This result is based on the "law of large numbers" and, while the rate of convergence is painfully slow, there are minimal assumptions about the location of points making it an extremely useful result. We state this standard result for ease of reference.

Theorem 5.12 (Law of Large Numbers) *Let $v_1, v_2, \ldots, v_m$ be independent and identically distributed random variables whose probability density function is denoted by $\mu(v)$. Let*

$$I = \int_{-\infty}^{\infty} f(v)\mu(v)dv$$

exist. Then

$$\frac{1}{m} \sum_{i=1}^{m} f(v_i)$$

converges to I in probability with m.

For Corollary 5.9 we have taken

$$\mu(v) = \begin{cases} \frac{1}{T}, & 0 \le v \le T, \\ 0 & \text{otherwise} \end{cases}$$

in the theorem. It is worth noting that we could have modified the cost function $J(u)$ to obtain a more general result. If we take

$$J(u) = \int_0^\infty (L_t(u) - f(t))^2 \mu(t)dt + \int_0^\infty u^2(t)dt, \qquad (5.20)$$

most of the previous results still hold. However, there are problems that need to be resolved such as the nature of the optimal solution and how one interprets splines on an infinite interval. For example, if $\mu(t) = \lambda e^{-\lambda t}$, this would produce a spline-like smoothing function which would have some predictive value for future times. In fact, the lack of predictive value is one of the more serious drawbacks of polynomial spline approximation.

For Corollary 5.10, we appeal to the general theory of Riemann sums and use the following theorem which can be found in Davis's classical book [22].

Theorem 5.13 (Convergence of Riemann Sums) *Let $f(v)$ be continuous in $[a, b]$. Then*

$$\left| \int_a^b f(v)dv - h \sum_{i=1}^m f(a + kh) \right| \le (b - a)w \left(\frac{b - a}{m} \right),$$

where

$$w(\delta) = \max_{v_1 - v_2 \le \delta} |f(v_1) - f(v_2)|, \quad a \le v_1, v_2 \le b.$$

There are various refinements of this theorem, and we refer the reader to [23] for a survey and interpretation of the literature. For this result, the rate of convergence is of the order of $\frac{1}{m}$ which is an improvement over the rate of convergence given by the law of large numbers which is only $\frac{1}{\sqrt{m}}$. There are various improvements which can be made along this line. For example we can use multiple point trapezoidal rules and multiple point Simpson's rules to obtain polynomial convergence of various orders. Again we refer the reader to [23] for many examples.

The results for Gaussian quadrature are quite diverse. Technically, we have used the following theorem related to Legendre quadrature.

Theorem 5.14 (Legendre Quadrature) *Let*

$$E_m(f) = \int_0^T f(v)dv - \sum_{k=1}^m w_k f(v_k),$$

where the v_k are the zeros of the Legendre polynomial of degree m and the weights w_k are the weights of the associated quadrature scheme. Then

$$E_m(f) = \frac{2^{2m+1}(m!)^4}{(2m+1)[(2m)!]^3} f^{(2m)}(\eta).$$

From this result we see that the order of convergence is nonpolynomial. We also see that it becomes harder and harder to give precise estimates because of the difficulty of estimating the higher derivatives of f. In [23] there are numerous results on the rates of convergence of some of the classical quadrature schemes, but there does not seem to be a general procedure for finding rates for arbitrary weight functions.

5.4 RATE OF CONVERGENCE

In this section, we examine the problem of determining the rate of convergence of the optimal control u_N to the optimal control u^*. In general the problem of precise estimates is beyond the scope of this chapter, but we can determine rates for the important case of cubic smoothing splines.

The rate of convergence depends on ϵ and on

$$\int_0^T \|A_N(s)\|^2 ds,$$

where

$$A_N(s)(u) = \sum_{i=1}^N w_{iN} \ell_{t_{iN}}(s) L_{t_{iN}}(u) = \int_0^T \sum_{i=1}^N w_{iN} \ell_{t_{iN}}(s) \ell_{t_{iN}}(r) u(r) dr.$$

We then calculate the norm of the linear functional

$$\|A_N(s)\| = \left(\int_0^T \left(\sum_{i=1}^N w_{iN} \ell_{t_{iN}}(s) \ell_{t_{iN}}(r) \right)^2 dr \right)^{\frac{1}{2}},$$

and hence

$$\int_0^T \|A_N(s)\|^2 ds = \int_0^T \int_0^T \left(\sum_{i=1}^N w_{iN} \ell_{t_{iN}}(s) \ell_{t_{iN}}(r) \right)^2 dr ds. \quad (5.21)$$

We now use the assumption that the sequence w_{iN}, t_{iN} defines a convergent quadrature scheme. We have

$$\left| \sum_{i=1}^N w_{iN} \ell_{t_{iN}}(s) \ell_{t_{iN}}(r) - \int_0^T \ell_t(s) \ell_t(r) dt \right| < \epsilon_N, \quad (5.22)$$

for some ϵ_N (dependent on the quadrature technique), and hence

$$\int_0^T \ell_t(s)\ell_t(r)dt - \epsilon_N < \sum_{i=1}^N w_{iN}\ell_{t_{iN}}(s)\ell_{t_{iN}}(r)$$

$$< \int_0^T \ell_t(s)\ell_t(r)dr + \epsilon_N.$$

However, we need estimates not on

$$\sum_{i=1}^N w_{iN}\ell_{t_{iN}}(s)\ell_{t_{iN}}(r)$$

but on

$$\left(\sum_{i=1}^N w_{iN}\ell_{t_{iN}}(s)\ell_{t_{iN}}(r)\right)^2.$$

This creates some problems since there is in general no reason to assume positivity.

5.5 CUBIC SPLINE CONVERGENCE BOUNDS

In this section, we restrict ourselves to the case that

$$A = \begin{pmatrix} 0 & 1 \\ 0 & 0 \end{pmatrix}, \quad b = \begin{pmatrix} 0 \\ 1 \end{pmatrix}, \quad c = \begin{pmatrix} 1 & 0 \end{pmatrix},$$

that is, to the case of cubic splines. We calculate

$$\ell_t(s) = \begin{cases} (t-s), & t > s, \\ 0 & \text{otherwise.} \end{cases}$$

This is usually denoted as $\ell_t(s) = (t-s)_+$ in the literature, and we will use this notation in this section. We further assume that the weights $w_{iN} > 0$. This is a minor assumption since all stable, convergent quadrature schemes have this property.

We choose ϵ_N sufficiently small that $\int_0^T \ell_t(s)\ell_t(r)dt - \epsilon_N$ is positive, and we have

$$\left(\int_0^T (t-s)_+(t-r)_+dt - \epsilon_N\right)^2 < \left(\sum_{i=1}^N (t_{iN}-s)_+(t_{iN}-r)_+\right)^2$$

$$< \left(\int_0^T (t-s)_+(t-r)_+dt + \epsilon_N\right)^2.$$

Now, by integrating and using (5.21), we have

$$\int_0^T \int_0^T \left(\int_0^T (t-s)_+(t-r)_+ dt - \epsilon_N \right)^2 dr ds \le \int_0^T \|A(s)\|^2 ds,$$

which in turn is less than or equal to

$$\int_0^T \int_0^T \left(\int_0^T (t-s)_+(t-r)_+ dt + \epsilon_N \right)^2 dr ds.$$

We now evaluate the integrals

$$\int_0^T \int_0^T \int_0^T (t-s)_+(t-r)_+ dt dr ds$$

and

$$\int_0^T \int_0^T \left(\int_0^T (t-s)_+(t-r)_+ dt \right)^2 dr ds.$$

We first assume that $r > s$ and we have

$$\int_0^T \int_0^T \int_0^T (t-s)_+(t-r)_+ dt dr ds$$
$$= \int_0^T \int_s^T \int_r^T (t-s)(t-r) dt dr ds.$$

This integral is tedious but routine to evaluate. In fact, we have

$$\int_0^T \int_0^T \int_0^T (t-s)_+(t-r)_+ dt dr ds = \frac{1}{40} T^5$$

and

$$\int_0^T \int_0^T \left(\int_0^T (t-s)_+(t-r)_+ dt \right)^2 dr ds = \frac{11}{3360} T^8.$$

Combining these two integrals we have

$$\frac{11}{3360} T^8 - \frac{1}{20} T^5 \epsilon_N + \frac{1}{2} T^2 \epsilon_N^2 \le \int_0^T \|A_N(s)\|^2 ds$$
$$\le \frac{11}{3360} T^8 + \frac{1}{20} T^5 \epsilon_N + \frac{1}{2} T^2 \epsilon_N^2.$$

So, we have

$$\|u_N - u^*\| \leq \frac{\epsilon_N}{\int_0^T \|A(s)\|^2 ds}$$

$$\leq \frac{\epsilon_N}{\frac{11}{3360}T^8 - \frac{1}{20}T^5\epsilon_N + \frac{1}{2}T^2\epsilon_N^2}$$

$$< \frac{\epsilon_N}{\frac{11}{3360}T^8 - \frac{1}{20}T^5\epsilon_N}.$$

The rate of the convergence of u_N to u^* is asymptotically the same as the rate of convergence of the quadrature scheme. This is the rate that would be expected for the "law of large numbers."

SUMMARY

In this chapter we showed that the smoothing spline is convergent in the L_2-sense, subject to certain minor assumptions. Moreover, the particular quadrature schemes of uniform Riemannian sums and Gaussian quadratures were shown to satisfy the necessary assumptions.

As a final note, the rate of convergence was characterized for cubic smoothing splines, where it was found that the convergence rate is asymptotically given by the rate of convergence of the quadrature scheme.

Chapter Six

SMOOTHING SPLINES WITH CONTINUOUS DATA

In the previous chapters we developed machinery for constructing smoothing splines with discrete data–both deterministic and random. However, there are many problems for which the data are continuous. EKG, ECG, and EMG are primary examples for which there is a continuous data stream, and, as shown in the previous chapter, smoothing splines for discrete data converge to splines with continuous data as the number of data points grows. In this chapter, we will develop a corresponding Hilbert space approach for smoothing continuous data with and without additional deterministic discrete data.

To establish the technique, we will solve the linear quadratic regulator problem and then use the construction for the smoothing problems. This construction is basically the same as is given in [28]. We are given a cost function

$$J(u) = \int_0^T \left[x^T(t)Qx(t) + u^2(t) \right] dt$$

and a controllable linear system

$$\dot{x} = Ax + bu, \quad x(0) = x_0,$$

and we assume that x_0 is given. We also assume that the matrix Q is positive definite. We define a Hilbert space

$$\mathcal{H} = L_2[0, T] \times L_2^n[0, T],$$

with norm

$$\|(u; x)\|_{\mathcal{H}}^2 = \int_0^T \left[x^T(t)Qx(t) + u^2(t) \right] dt.$$

Let the constraint variety be defined as

$$V_{x_0} = \left\{ (u; x) \; \middle| \; x(t) = e^{At}x_0 + \int_0^t e^{A(t-s)}bu(s)ds \right\}.$$

Note again that V_{x_0} (and hence also V_0) is closed by the closed graph theorem. As for the discrete case, we can minimize the cost function by finding the point of minimum norm in V_{x_0}. Thus we construct the orthogonal complement of V_0. We have

$$V_0^\perp = \left\{ (v; w) \,\middle|\, \int_0^T \left[x^T(t) Q w(t) + u(t) v(t) \right] dt = 0, \ \forall (u; x) \in V_0 \right\}.$$

Using this definition, we have

$$0 = \int_0^T \left[x^T(t) Q w(t) + u(t) v(t) \right] dt$$

$$= \int_0^T \left[w^T(t) Q \int_0^t e^{A(t-s)} b u(s) ds + u(t) v(t) \right] dt$$

$$= \int_0^T \int_s^T w^T(t) Q e^{A(t-s)} b u(s) dt ds + \int_0^T u(s) v(s) ds$$

$$= \int_0^T \left[\int_s^T w^T(t) Q e^{A(t-s)} b \, dt + v(s) \right] u(s) ds.$$

Thus we have

$$V_0^\perp = \left\{ (v; w) \,\middle|\, v(s) = - \int_s^T b^T e^{A^T(t-s)} Q w(t) dt \right\}.$$

In order to find the intersection $V_{x_0} \cap V_0^\perp$ we must solve the following system of two integral equations:

$$x(t) = e^{At} x_0 + \int_0^t e^{A(t-s)} b u(s) ds, \tag{6.1}$$

$$u(s) = - \int_s^T b^T e^{A^T(t-s)} Q x(t) dt. \tag{6.2}$$

To solve this system, we let $u(t) = -b^T \lambda(t)$. From (6.2) we have

$$\lambda(t) = \int_t^T e^{A^T(r-t)} Q x(r) dr,$$

and from (6.1) we have

$$x(t) = e^{At} x_0 - \int_0^t e^{A(t-s)} b b^T \lambda(s) ds.$$

Differentiating these two equations, we have the standard Hamiltonian formulation of the optimal control problem:

$$\frac{d}{dt} \begin{bmatrix} x(t) \\ \lambda(t) \end{bmatrix} = \begin{bmatrix} A & -bb^T \\ -Q & -A^T \end{bmatrix} \begin{bmatrix} x(t) \\ \lambda(t) \end{bmatrix}, \quad \begin{bmatrix} x(0) \\ \lambda(T) \end{bmatrix} = \begin{bmatrix} x_0 \\ 0 \end{bmatrix}. \tag{6.3}$$

The solution of the equation is then done by introducing the Riccati transform. (See, for example, [28], or any introductory control text.)

6.1 CONTINUOUS DATA

Continuous data arise, for example, from the output of analog devices. While we usually assume that such devices are accurate, any observation of the analog output of a six-lead electrocardiograph over a period of an hour or so should convince even the most causal observer that the devices are not exact and are subject to considerable error of a random nature.

The classic problem of this nature is that of determining the bottom profile of a lake. A depth finder is employed and a trace is made as a boat traverses the lake. Chesser [16] has posed the problem of determining the radioactivity of the silt at the bottom of the cooling pond at the reactor site in Chernobyl. A device would be carried along the bottom measuring radioactivity as a continuous function. Such data would be naturally very stochastic, and a major problem is to smooth the continuous data so that a reliable map of the radioactivity can be obtained.

Data such as these also arise in certain cryptological applications. For example, an encoded signal can be attached to the output of an FM radio station. Here the object is not to recover the radio signal but to recover the static. Thus if the received signal is recorded and smoothed it can be subtracted from the received signal and an approximation of the static recovered. The static can then be treated with classical cryptological methods to recover the message. This problem was studied in [59].

After this rather informal discussion about the nature of continuous data, we are ready to formulate the basic problem.

6.2 THE CONTINUOUS SMOOTHING PROBLEM

As before, we will assume a model of the form

$$\dot{x} = Ax + bu, \quad y = cx, \tag{6.4}$$

with boundary data given by

$$L_0 x(0) + \sum_{i=1}^{N-1} L_i x(t_i) + L_N x(T) = \gamma. \tag{6.5}$$

We make the usual assumptions that $x \in \mathbb{R}^n$ and that L_i is a map from $\mathbb{R}^n$ to $\mathbb{R}^k$. As for the boundary data, we assume that there is at least one solution to (6.5), and we will assume the following consistency condition:

$$\text{If, for every } i, a^T L_i = 0 \text{ then } a^T = 0.$$

The data will be assumed to be a square integrable function on the interval $[0, T]$, that is, $f \in L_2[0, T]$. Following the general algorithm of the previous chapter, we define a Hilbert space $\mathcal{H}$ that (with some minor modifications) will underlie the rest of this chapter. Let

$$\mathcal{H} = L_2[0, T] \times L_2[0, T] \times \mathbb{R}^{n(N+1)}, \tag{6.6}$$

where we have

$$(u; y(t); x(0); x(t_1); \ldots; x(t_{N-1}), x(T)) \in \mathcal{H},$$

with norm

$$\|(u; y(t); x(0); x(t_1); \ldots; x(t_{N-1}), x(T)\|_{\mathcal{H}}$$

being equal to

$$\int_0^T \left[\lambda_1 u^2(t) + \lambda_2 y^2(t) \right] dt + \hat{x}^T Q \hat{x},$$

where $\hat{x} = (x(0)^T, x(t_1)^T, \ldots, x(t_{N-1})^T, x(T)^T)^T$, Q is positive definite, and the λ_i are positive.

The constraints, defined through the boundary data and the system dynamics, then generate an affine variety in $\mathcal{H}$. Let

$$V_\gamma = \Big\{ ((u; y(t); x(0); x(t_1); \ldots; x(t_{N-1}); x(T)) \in \mathcal{H} \;\Big|$$

$$y(t) = ce^{At}x_0 + \int_0^t ce^{A(t-s)}bu(s)ds,$$

$$L_0 x(0) + L_N x(T) + \sum_{i=1}^{N-1} L_i x(t_i) = \gamma \Big\}.$$

We first note that under the assumptions we have made the variety is non-empty. We have assumed that there exists at least one solution to the equation defining the boundary conditions and thus, by controllability, there exists a control that drives the solution through those points. Thus the conditions are satisfied for at least one point.

Remark: We identify the data with the point

$$(0; f; 0; \ldots; 0) \in \mathcal{H}.$$

Now we have a welldefined problem of finding the unique point in the affine variety V_γ that is nearest to the point representing the data. It is worth noting that we have defined the constraint in terms of an initial value problem. Both Willsky and Krener worked directly with the boundary value problem, but this is really unnecessary in this context. They were assuming a known dynamical model that generates the data. We are not assuming any such model, but are only approximating and smoothing the data using a model to generate the approximating data. The approach we are using generalizes the approach used in Chapter 3. There the Hilbert space that contains the data was finite-dimensional; here it is not.

6.3 THE BASIC TWO-POINT BOUNDARY VALUE PROBLEM

In this section we will consider the simplest set of boundary conditions:

$$L_1 = [I, 0], \quad L_2 = [0, I].$$

We then have

$$[x(0), x(T)] = L_1 x(0) + L_2 x(T) = \gamma = [\alpha, \beta].$$

This setup is just the point-to-point transfer problem, albeit with an associated target function, f.

The affine variety we that we will use is

$$V_\gamma = \left\{ (u; y) \mid e^{-AT} \beta - \alpha = \int_0^T e^{-As} bu(s) ds, \right.$$

$$\left. y(t) = ce^{At} \alpha + \int_0^t ce^{A(t-s)} bu(s) ds \right\},$$

and the Hilbert space is

$$\mathcal{H} = L_2[0, T] \times L_2[0, T].$$

The goal is to find $y(t)$ that best approximates the data function

$$f \in L_2[0, T].$$

Thus we look for the point in V_γ that is closest to the point

$$(0, f) \in \mathcal{H}.$$

We already have stated repeatedly that this point is constructed by first constructing the subspace $V_0^\perp$, and then constructing the intersection of

$$V_\gamma \cap (V_0^\perp + (0, f)).$$

This intersection consists of a single point, provided that V_0 is closed. But the operators that define V_γ are continuous and hence the space is closed.

Note that we can rewrite V_γ as

$$\left\{ \left(u; ce^{At}\alpha + \int_0^t ce^{A(t-s)}bu(s)ds \right) \,\middle|\, e^{-AT}\beta - \alpha = \int_0^T e^{-As}bu(s)ds \right\}$$

and so V_γ is described as the graph of an operator with domain a linear variety.

Now, we have

$$V_0 = \left\{ \left(u; \int_0^t ce^{A(t-s)}bu(s)ds \right) \,\middle|\, 0 = \int_0^T e^{-As}bu(s)ds \right\},$$

and

$$V_0^{\perp} = \left\{ (v; w) \,\middle|\, \int_0^T v(s)u(s)ds + \int_0^T w(t)\int_0^t ce^{A(t-s)}bu(s)dsdt = 0, \right.$$
$$\left. 0 = \int_0^T e^{-As}bu(s)ds \right\}.$$

Using the calculations employed in previous chapters, we get

$$\int_0^T \left(v(s) + \int_s^T w(t)ce^{A(t-s)}bdt \right) u(s)ds = 0.$$

Normally, we would have concluded that

$$v(s) + \int_s^T w(t)ce^{A(t-s)}bdt = 0,$$

but since u is restricted, we can only conclude that

$$v(s) + \int_s^T ce^{A(t-s)}bw(t)dt = \lambda^T e^{-As}b.$$

Thus

$$V_0^{\perp} = \left\{ (v; w) \,\middle|\, v(s) + \int_s^T ce^{A(t-s)}bw(t)dt = \lambda^T e^{-As}b, \ \lambda \in \mathbb{R}^n \right\},$$
$$\tag{6.7}$$

and we state this as a lemma.

Lemma 6.1 *Let*

$$\mathcal{H} = L_2[0, T] \times L_2[0, T],$$

and let V_γ be given by

$$\left\{ \left(u; ce^{At}\alpha + \int_0^t ce^{A(t-s)}bu(s)ds \right) \;\middle|\; e^{-AT}\beta - \alpha = \int_0^T e^{-As}bu(s)ds \right\}.$$

Then

$$V_0^\perp = \left\{ (v; w) \;\middle|\; v(s) + \int_s^T ce^{A(t-s)}bw(t)dt = \lambda^T e^{-As}b, \; \lambda \in \mathbb{R}^n \right\}.$$

We now construct the intersection. First, consider $V_0^\perp + (0; f)$ and note that we must have $w + f = y$ and hence $w = y - f$. Therefore, let

$$\eta(t) = \int_t^T e^{-A^T(t-s)}c^T(y(s) - f(s))ds. \tag{6.8}$$

Thus we have

$$\dot{\eta}(t) = -c^T(y(t) - f(t)) - A^T\eta(t) = -c^T cx(t) - A^T\eta(t) + c^T f(t) \tag{6.9}$$

and

$$\dot{x}(t) = Ax(t) - bb^T\eta(t) + bb^T e^{-A^T t}\lambda. \tag{6.10}$$

We, therefore, have the forced Hamiltonian system

$$\frac{d}{dt}\left(\begin{array}{c} x(t) \\ \eta(t) \end{array} \right) = \left(\begin{array}{cc} A & -bb^T \\ -c^T c & -A^T \end{array} \right)\left(\begin{array}{c} x(t) \\ \eta(t) \end{array} \right) + \left(\begin{array}{c} bb^T e^{-At}\lambda \\ c^T f(t) \end{array} \right), \tag{6.11}$$

with

$$x(0) = \alpha \quad \eta(T) = 0$$

and the added constraint

$$x(T) = \beta.$$

Since there exists a unique solution, we see that there will exist a one-to-one mapping from the terminal conditions to the λ and, for the same reason, a one-to-one mapping from initial data to the λ. Thus, we can solve the forced Hamiltonian system using variation of parameters.

Let

$$\left(\begin{array}{cc} X_1 & X_2 \\ X_3 & X_4 \end{array} \right) = \exp\left(\left(\begin{array}{cc} A & -bb^T \\ -c^T c & -A^T \end{array} \right)t \right),$$

$$\begin{pmatrix} x(t) \\ \eta(t) \end{pmatrix} = \begin{pmatrix} X_1(t-T) & X_2(t-T) \\ X_3(t-T) & X_4(t-T) \end{pmatrix} \begin{pmatrix} \beta \\ 0 \end{pmatrix}$$
$$- \int_t^T \begin{pmatrix} X_1(t-s) & X_2(t-s) \\ X_3(t-s) & X_4(t-s) \end{pmatrix} \begin{pmatrix} bb^T e^{-As}\lambda \\ c^T f(s) \end{pmatrix} ds.$$

It only remains to determine λ. However, we can uniquely solve the system of equations obtained for λ by setting $t = 0$. Solving, we have

$$x(t) = X_1(t_T)\beta - \int_t^T X_1(t-s)bb^T e^{-As}ds\lambda - \int_t^T X_2(t-s)c^T f(s)ds,$$

and setting $t = 0$ we have

$$\alpha = X_1(-T)\beta - \int_0^T X_1(-s)bb^T e^{-AS}ds\lambda - \int_0^T X_2(-s)c^T f(s)ds.$$

We have thus derived the following theorem.

Theorem 6.2 *The optimal approximation of the continuous data point, f, generated by the two-point boundary value problem*

$$\dot{x} = Ax + bu, \quad y = cx$$

with boundary data

$$x(0) = \alpha, \quad x(T) = \beta,$$

is given by $y(t)$ being equal to

$$cX_1(t-T)\beta - \int_t^T cX_1(t-s)bb^T e^{-As}ds\lambda - \int_t^T cX_2(t-s)c^T f(s)ds,$$

where λ is given by

$$\left(\int_0^T X_1(-s)bb^T e^{-As}ds \right)^{-1}$$
$$\times \left(-\alpha + X_1(-T)\beta - \int_0^T X_2(-s)c^T f(s)ds \right).$$

6.4 THE GENERAL TWO-POINT BOUNDARY VALUE PROBLEM

We now consider the problem when we have the more interesting boundary conditions

$$L_1 x(0) + L_2 x(T) = \gamma.$$

Recall that we work under the assumption that there exists at least one solution to the boundary equation. Let $x(0) = \alpha$ and $x(T) = \beta$ be one such solution. Because of the linearity, every solution $(x(0), x(T))$ can be written as

$$(x(0), x(T)) = (\alpha, \beta) + (\rho, \theta),$$

where

$$L_1 \rho + L_2 \theta = 0.$$

This problem includes the special case of controllability from point to line and from line to line. (These cases will be considered in detail in Chapter 9.)

If there is a unique solution to the boundary conditions then there is nothing left to do. If the solution is not unique, then we have extra degrees of freedom that must be used. We thus extend the Hilbert space to include the boundary conditions. Let

$$\mathcal{H} = L_2[0, T] \times L_2[0, T] \times \mathbb{R}^n \times \mathbb{R}^n.$$

We construct the affine variety as

$$V_\gamma = \Big\{ (u; y; x(0); x(T)) \;\Big|\; y(t) = ce^{At} x(0) + \int_0^t ce^{A(t-s)} bu(s)\,ds,$$

$$L_1 x(0) + L_2 x(T) = \gamma,\; e^{-AT} x(T) - x(0) = \int_0^T e^{-As} bu(s)\,ds \Big\}.$$

As before, we must construct $V_0^\perp$, and as in the previous section, we have that $V_0^\perp$ is given by

$$\Big\{ (v; w; \tau; \phi) \;\Big|\; \int_0^T [v(t)u(t) + w(t)y(t)]\,dt + \tau^T x(0) + \psi^T x(T) = 0 \Big\}.$$

Substituting the definition of y into the expression gives

$$0 = \int_0^T \left[v(s) + \int_s^T ce^{A(t-s)} bw(t)\,dt \right] u(s)\,ds$$

$$+ \int_0^T ce^{At} x(0) w(t)\,dt + \tau^T x(0) + \psi^T x(T).$$

We can break this into two parts, since it must be true for $u = 0$ and $x(0)$ and $x(T)$ arbitrary solutions of the boundary equations, and, likewise, it must be true for u arbitrary with 0 boundary conditions. Thus, as in the previous section, we conclude that

$$v(s) + \int_s^T ce^{A(t-s)}bw(t)dt = b^T e^{-A^T s}\lambda.$$

From

$$\int_0^T ce^{At}x(0)w(t)dt + \tau^T x(0) + \psi^T x(T) = 0,$$

we conclude that

$$\left[\int_0^T ce^{At}w(t)dt + \tau^T\right]x(0) + (\psi^T)x(T) = 0,$$

and hence that there exists a vector k such that

$$k^T L_1 = \int_0^T ce^{At}w(t)dt + \tau^T$$

and

$$k^T L_2 = \psi^T.$$

We then have the following lemma.

Lemma 6.3 *Let $\mathcal{H} = L_2[0,T] \times L_2[0,T] \times \mathbb{R}^n \times \mathbb{R}^n$ and let*

$$V_\gamma = \left\{ (u; y; x(0); x(T)) \;\middle|\; y(t) = ce^{At}x(0) + \int_0^t ce^{A(t-s)}bu(s)ds, \right.$$

$$\left. L_1 x(0) + L_2 x(T) = \gamma, \; e^{-AT}x(T) - x(0) = \int_0^T e^{-As}bu(s)ds \right\}.$$

Then $V_0^\perp$ is given by the set

$$\left\{ \left(-\int_s^T ce^{A(t-s)}bw(t)dt + b^T e^{-A^T s}\lambda; w(t); \right.\right.$$

$$\left.\left. k^T L_1 - \int_0^T ce^{At}w(t)dt; k^T L_2 \right) \right\},$$

where $w \in L_2[0,T]$, $\lambda \in \mathbb{R}^n$, $k \in \mathbb{R}^k$.

We now construct $(V_0^\perp + (0; f; 0; 0)) \cap V_\gamma$. Exactly as in the previous section, we have the forced terminal value problem

$$\frac{d}{dt}\begin{pmatrix} x(t) \\ \eta(t) \end{pmatrix} = \begin{pmatrix} A & -bb^T \\ -c^T c & -A^T \end{pmatrix}\begin{pmatrix} x(t) \\ \eta(t) \end{pmatrix} + \begin{pmatrix} bb^T e^{-At}\lambda \\ c^T f(t) \end{pmatrix},$$

with

$$x(T) = L_2^T k, \ \eta(T) = 0.$$

Here, both k and λ are parameters to be determined. Using the notation from the previous section, we have

$$x(t) = X_1(t-T)L_2^T k - \int_t^T X_1(t-s)bb^T e^{-As} ds\lambda - \int_t^T X_2(t-s)c^T f(s)ds$$

and

$$x(0) = X_1(-T)L_2^T k - \int_0^T X_1(-s)bb^T e^{-As} ds\lambda - \int_0^T X_2(-s)c^T f(s)ds.$$

We now have to determine the values of k and λ. We need $n + m$ equations. We obtain m equations from the boundary conditions

$$\gamma = L_1(X_1(-T)L_2^T k - \int_0^T X_1(-s)bb^T e^{-As} ds\lambda$$
$$- \int_0^T X_2(-s)c^T f(s)ds) + L_2 L_2^T k,$$

and n equations from

$$L_1^T k - \int_0^T e^{A^T t}c^T w(t)dt = X_1(-T)L_2^T k - \int_0^T X_1(-s)bb^T e^{-As} ds\lambda$$
$$- \int_0^T X_2(-s)c^T f(s)ds.$$

Unfortunately this equation contains w.

Recalling that $w = y - f$ and substituting, we have

$$L_1^T k - \int_0^T e^{A^T t}c^T y(t)dt + \int_0^T e^{A^T t}c^T f(t)dt$$
$$= X_1(-T)L_2^T k - \int_0^T X_1(-s)bb^T e^{-As} ds\lambda - \int_0^T X_2(-s)c^T f(s)ds.$$

Now, $y(t)$ becomes

$$cX_1(t-T)L_2^T k - \int_t^T cX_1(t-s)bb^T e^{-As} ds\lambda - \int_t^T cX_2(t-s)c^T f(s)ds,$$

which, after some manipulation, yields

$$\left(L_1^T - \int_0^T e^{A^T t} c^T c X_1(t-T) L_2^T dt - X_1(-T) L_2^T \right) k$$

$$+ \left(\int_0^T \int_t^T e^{A^T t} c^T c X_1(t-s) b b^T e^{-As} ds dt \right.$$

$$\left. - \int_0^T X_1(-s) b b^T e^{-As} ds \right) \lambda$$

$$= - \int_0^T \int_t^T e^{A^T t} c^T c X_2(t-s) c^T f(s) ds dt - \int_0^T X_2(-s) c^T f(s) ds$$

$$- \int_0^T e^{A^T t} c^T f(t) dt.$$

We have thus established the following theorem.

Theorem 6.4 *The optimal approximation of the continuous data point, f, generated by the two-point boundary value problem*

$$\dot{x} = Ax + bu, \quad y = cx,$$

with boundary data

$$L_1 x(0) + L_2 x(T) = \gamma$$

is given by

$$y(t) = c X_1(t-T) L_2^T k - \int_t^T c X_1(t-s) b b^T e^{-As} ds \lambda$$

$$- \int_t^T c X_2(t-s) c^T f(s) ds,$$

where k and λ are the unique solutions of

$$\left(L_1^T - \int_0^T e^{A^T t} c^T c X_1(t-T) L_2^T dt - X_1(-T) L_2^T \right) k$$

$$+ \left(\int_0^T \int_t^T e^{A^T t} c^T c X_1(t-s) b b^T e^{-As} ds dt \right.$$

$$\left. - \int_0^T X_1(-s) b b^T e^{-As} ds \right) \lambda$$

$$= - \int_0^T \int_t^T e^{A^T t} c^T c X_2(t-s) c^T f(s) ds dt - \int_0^T X_2(-s) c^T f(s) ds$$

$$- \int_0^T e^{A^T t} c^T f(t) dt,$$

and

$$(L_1(X_1(-T)L_2^T + l_2 L_2^T)k - \left(\int_0^T X_1(-s)bb^T e^{-As} ds \right) \lambda$$
$$= \int_0^T X_2(-s)c^T f(s)ds) + \gamma.$$

6.5 MULTIPOINT PROBLEMS

In this section we will examine the case when there are boundary points interior to the interval. While the general approach is the same as for the two-point boundary value problems, there are technical difficulties. Problems such as this arise in map building where there are some known points of reference. For example, when constructing topographical maps some points are determined by physical survey. These points are assumed to be precisely known and must be respected by any path planning algorithm.

As before, we assume as given the standard linear single input single output system, with boundary conditions given by

$$L_0 x(0) + \sum_{i=1}^{N-1} L_i x(t_i) + L_N x(T) = \gamma.$$

We assume that there is at least one solution respecting these boundary conditions. We assume that the data are given by a square integrable function $f \in L_2[0, T]$.

We define a Hilbert space $\mathcal{H}$ as

$$\mathcal{H} = L_2[0, T] \times L_2[0, T] \times \mathbb{R}^{n(N+1)},$$

with norm given by

$$\| (u; y; x(0); x(T_1); \ldots; x(T_{N-1}); x(T)) \|_{\mathcal{H}}^2$$
$$= \int_0^T (\lambda_1 u^2(t) + \lambda_2 y^2(t))dt + \eta^T Q \eta,$$

where

$$\eta = (x(0)^T; x(T_1)^T; \ldots; x(T_{N-1})^T; x(T)^T)^T$$

and Q is positive definite.

Let

$$L = [L_0, L_1, \ldots, L_{N-1}, L_N].$$

We can then define the linear variety of constraints as

$$V_\gamma = \left\{ (u; y; \eta) \in \mathcal{H} \;\middle|\; y(t) = ce^{At}x(0) + \int_0^t ce^{A(t-s)}bu(s)ds; \; L\eta = \gamma \right\}.$$

Because we have assumed that there is at least one solution to the boundary constraints, the variety is nonempty. As we saw in Section 6.3, we can replace the boundary condition with the simpler condition

$$\left(\sum_{i=0}^N L_i e^{AT_i} \right) x(0) + \int_0^T \left(\sum_{i=1}^N L_i G_i(s) \right) u(s)ds = \gamma.$$

Thus we can rewrite V_γ in terms of $x(0)$, $y(t)$, and $u(t)$, and can pose the problem in a simpler Hilbert space. We define a new Hilbert space $\mathcal{W}$ as

$$\mathcal{W} = L_2[0,T] \times L_2[0,T] \times \mathbb{R}^n,$$

with norm

$$\|(u; y; x)\|_{\mathcal{W}}^2 = \int_0^T (\lambda_1 u^2(t) + \lambda_2 y^2(t))dt + x^T Q x,$$

where Q is a positive definite $n \times n$ matrix. We can then rewrite the variety V_γ as

$$V_\gamma = \left\{ (u; y; x(0)) \in \mathcal{W} \;\middle|\; y(t) = ce^{At}x(0) + \int_0^t ce^{A(t-s)}bu(s)ds, \right.$$
$$\left. \left(\sum_{i=0}^N L_i e^{AT_i} \right) x(0) + \int_0^T \left(\sum_{i=1}^N L_i G_i(s) \right) u(s)ds = \gamma \right\}.$$

We now construct $V_0^\perp$. By definition, we have

$$V_0^\perp = \left\{ (v; w; z) \;\middle|\; \forall \, (u; y; x(0)) \in V_0, \int_0^T (\lambda_1 v(t)u(t) + \lambda_2 y(t)w(t))dt \right.$$
$$\left. + z^T Q x(0) = 0 \right\}.$$

From the definition of V_0, we can simplify the condition in the orthogonal complement by requiring y to satisfy the following expression:

$$0 = \int_0^T \left[\lambda_1 v(t)u(t) + \lambda_2 \left(ce^{At}x(0) + \int_0^t ce^{A(t-s)}bu(s)ds \right) w(t) \right] dt$$
$$+ z^T Q x(0).$$

Rewriting this after some manipulation involving a change of the order of integration, we have

$$0 = \int_0^T \left[\lambda_1 v(s) + \int_s^T \lambda_2 c e^{A(t-s)} b w(t) dt \right] u(s) ds$$

$$\left[\lambda_2 \int_0^T c e^{At} w(t) dt + z^T Q \right] x(0) = 0.$$

This expression holds for all $x(0)$ and u, and so we have

$$\lambda_1 v(s) + \int_s^T \lambda_2 c e^{A(t-s)} b w(t) dt = 0$$

and

$$\int_0^T \lambda_2 c e^{At} w(t) dt + z^T Q = 0.$$

Thus we have the simpler expression for the complement,

$$V_0^{\perp} = \left\{ (v; w; z) \ \middle| \ \lambda_1 v(s) + \int_s^T \lambda_2 c e^{A(t-s)} b w(t) dt = 0, \right.$$

$$\left. \int_0^T \lambda_2 c e^{At} w(t) dt + z^T Q = 0 \right\}.$$

Now the data we seek to approximate are a function $f \in L_2[0,T]$. So we seek the point on V_γ nearest in the sense of the norm to $(0; f; 0)$. As in [64], this point is found by finding the intersection of $V_0^{\perp} \cap V_\gamma$. There is a unique point of intersection, and the construction is basically the same as in the previous sections. We leave the construction to the reader.

As an application of splines with continuous data, we consider a problem involving the recursive generation of splines. The basic idea is to encode past data as a spline and then use this curve (continuous data) together with new data in order to recursively update the spline. The update is approximated using a quadrature method that makes the problem computationally feasible.

6.6 RECURSIVE SPLINES

Splines, both smoothing and interpolating, are ubiquitous in all problems in which it is required to construct a curve from data. However, for large data sets, the direct methods of construction involve solving large systems of linear equations and/or inverting large matrices. In this section, we investigate the problem of treating large data sets in a recursive manner in order to keep the dimensions of problems to be solved under a fixed size.

As in previous sections in this chapter, the particular problem that motivated this work was boundary reconstruction when the boundary is being measured at isolated points using a remote device. It is assumed that the boundary is closed and that arbitrarily many measurements can be made. It is, moreover, assumed that the device can make a series of measurements in one complete revolution and that additional revolutions could made. Thus we assume that N measurement are made at each revolution and a smoothing spline is constructed after the first revolution is complete. After the second revolution, the second set of data is to be used to modify the first smoothing spline, and so on. The idea is very similar to the problem of using new data to update an existing map. (Results of this problem have been reported in [54].)

Accordingly, we assume a sequence of data sets of equal size,

$$D_n = \{\alpha_{in} : i = 1, \dots, k\}. \tag{6.12}$$

We further assume that the data are of the form

$$\alpha_{in} = f(t_{in}) + \epsilon_{in},$$

where $f(t)$ is a continuous function that is at least piecewise smooth and the ϵ_{in} are values of an iid random variable that is at least symmetrically distributed about 0. These conditions have been studied in [32] and [104], and also in Chapter 5.

We assume that the set $\{t_{in} : i = 1, \dots, k, \ n = 1, 2, \dots\}$ is dense in the interval $[0, T]$. For $n = 1, 2, \dots$ we let the cost function be given by

$$J_n(u) = \int_0^T u(t)^2 dt + \lambda_n \int_0^T (y(t) - y_{n-1}(t))^2 dt + \sum_{i=1}^k (y(t_{in}) - \alpha_{in})^2.$$

We let $u_n(t)$ and $y_n(t)$ be the optimal control and resulting output with respect to this cost function. Here the idea is that we are encoding the past data as the spline $y_{n-1}(t)$. The coefficients λ_n form a sequence of numbers that approach infinity. One of the main goals of this section is to show that this sequence can be chosen in such a way that the sequence of smoothing splines $\{y_n(t)\}_{n=1}^\infty$ converges.

In [32] and Chapter 5 we studied the following problem. Let

$$J^N(u) = \int_0^T u(t)^2 dt + \sum_{n=1}^N \sum_{i=1}^k (y(t_{in}) - \alpha_{in})^2. \tag{6.13}$$

There we showed that the optimal control and splines converge under mild assumptions on the data, mainly that the ϵ_{in} are symmetrically distributed and that first and second moments exist. The problem with this approach is that the linear systems that must be solved grow without bound, creating insurmountable numerical difficulties.

To circumvent this problem we return to the Hilbert space technique, and formulate the problem as a minimum norm problem in a particular Hilbert space. We follow closely the development in the previous section. We let

$$\mathcal{H}_n = \{(u; g; \alpha) \mid (u; g; \alpha) \in L_2[0, T] \times L_2[0, T] \times \mathbb{R}^k\},$$

with norm

$$\|(u; g; \alpha)\|_{\mathcal{H}_n}^2 = \int_0^T u(t)^2 dt + \lambda_n \int_0^T g(t)^2 dt + \alpha^T \alpha.$$

As before, let

$$\ell_{in}(s) = \begin{cases} ce^{A(t_{in}-s)}b, & t_{in} - s \geq 0, \\ 0 & \text{otherwise.} \end{cases}$$

We now define a linear affine variety

$$V_{x_0} = \Big\{(u; y; z) \mid y(t) = ce^{At}x_0 + \int_0^T ce^{A(t-s)}bu(s)ds,$$

$$z_i = ce^{At_{in}}x_0 + \int_0^T \ell_{in}(s)u(s)ds\Big\}. \tag{6.14}$$

We define the data point in $\mathcal{H}_n$ to be the point

$$p_n = (0, y_{n-1}(t), \alpha^n), \tag{6.15}$$

where

$$\alpha^n = (\alpha_{1n}, \ldots, \alpha_{kn}).$$

The optimization problem is now to find the unique point in the linear variety V_{x_0} that is closest to the data point p_n. However, since the data are continuous this results in solving a system of linear integral equations and effort reduction is lost. The basic question is "Does there exist a sequence of λ_n so that the solutions converge to the underlying the function $f(t)$?" The trick to answering this question is to recast the problem as one with only discrete data. This simplification greatly reduces the complexity of the problem. We approximate the cost function $J_n(u)$ with a new cost function $J_n^j(u)$, by replacing the second integral with a finite sum.

A quadrature scheme is defined by two data sets given by two real lower triangular matrices whose jth rows are given by

$$T_j = (r_{j1}, \ldots, r_{jj}, 0, 0, \ldots)$$

and

$$S_j = (\beta_{j1}, \ldots, \beta_{jj}, 0, 0, \ldots).$$

Let f be any continuous function, and let

$$E_j(f) = \left| \int_0^T f(t)dt - \sum_{i=1}^j \beta_{ji} f(r_{ji}) \right|.$$

The quadrature scheme is convergent if, for every $f \in C[0, T]$, we have

$$\lim_{j \to \infty} E_j(f) = 0.$$

Such schemes abound (see, for example, [23]), and we choose any such convergent scheme.

We now define the new cost function

$$J_n^j(u) = \int_0^T u(t)^2 dt + \lambda_n \sum_{i=1}^j \beta_{ij} (y(r_{ij}) - y_{n-1}(r_{ij}))^2 + \sum_{i=1}^k (y(t_{in}) - \alpha_{in})^2$$

(6.16)

in terms of the convergent quadrature scheme. We again rephrase the optimization problem as a minimum norm problem in Hilbert space. Here the past data are encoded in the spline function, but we are only taking particular snapshots of the spline. The relationship between these two problems clearly depends on the accuracy of the chosen quadrature scheme. However, using the methods of [32] and [104] it can be established that the solutions of J_n^j converge to those of J_n as j approaches infinity.

6.6.1 The Continuous Case

Let

$$V_{x_0}^k = \left\{ (u; y; y(t_{ik})) \,\middle|\, y(t) = ce^{At}x_0 + \int_0^t ce^{A(t-s)}bu(s)ds, \right.$$

$$\left. y(t_{ik}) = ce^{At_i}x_0 + \int_0^T \ell_{ik}(s)u(s)ds \right\}.$$

Then

$$V_0^k = \left\{ (u; y; y(t_{ik})) \,\middle|\, y(t) = \int_0^t ce^{A(t-s)}bu(s)ds, \right.$$

$$\left. y(t_{ik}) = \int_0^T \ell_{ik}(s)u(s)ds \right\}$$

and $V^{\perp}$ is given by

$$\left\{(w; z; \beta) \;\middle|\; \int_0^T w(s)u(s)ds + \lambda_k \int_0^T z(t)y(t)dt + \sum_{i=1}^N y(t_{ik})\beta_i = 0\right\},$$

from which we conclude that

$$V^{\perp} = \left\{(w; z; \beta) \;\middle|\; w(s) + \lambda_k \int_s^T z(t)ce^{A(t-s)}bdt + \sum_{i=1}^N \ell_{ik}(s)\beta_i = 0\right\}.$$

Constructing the intersection of $V_{x_0}^k$ with $V^{\perp} + p$, we must solve the following system of equations:

$$u(s) = -\lambda_k \int_s^T (y(t) - y_n(t))ce^{A(t-s)}bdt - \sum_{i=1}^N \ell_{ik}(s)(y(t_{ik}) - \alpha_{ik}),$$

$$\tag{6.17}$$

$$y(t) = ce^{At}x_0 + \int_0^t ce^{A(t-s)}bu(s)ds, \tag{6.18}$$

$$y(t_{jk}) = ce^{At_i}x_0 + \int_0^T \ell_{jk}(s)u(s)ds. \tag{6.19}$$

Substituting u into the other two equations, we have

$$y(t) = -\lambda_k \int_0^t \int_s^T y(t)ce^{2A(t-s)}bdtds - \sum_{i=1}^N \int_0^t ce^{A(t-s)}b\ell_{ik}(s)dsy(t_{ik})$$

$$+ \lambda_k \int_0^t \int_s^T y_n(t)ce^{A(t-s)}bce^{A(t-s)}bdtds$$

$$+ \sum_{i=1}^N \int_0^t ce^{A(t-s)}b\ell_{ik}(s)ds\alpha_{ik} + ce^{At}x_0,$$

$$y(t_{jk}) = -\lambda_k \int_0^T \ell_{jk}(s) \int_s^T y(t)ce^{A(t-s)}bdtds$$

$$- \sum_{i=1}^N \int_0^T \ell_{jk}(s)\ell_{ik}(s)dsy(t_{jk})$$

$$+ \lambda_k \int_0^T \ell_{jk}(s) \int_s^T y_n(t)ce^{A(t-s)}bdtds$$

$$+ \sum_{i=1}^N \int_0^T \ell_{jk}(s)\ell_{ik}(s)ds\alpha_{ik} + ce^{At_{jk}}x_0.$$

Let

$$L(t)(f) = \int_0^t \int_s^T f(t)ce^{2A(t-s)}bdtds, \tag{6.20}$$

and let

$$G_k = \left[\int_0^T \ell_{ik}(s)\ell_{jk}(s)ds \right]_{i,j=1}^N \tag{6.21}$$

be the Grammian of the set of linearly independent functions $\ell_i(s)$. Moreover, let

$$H_k = \left(\int_0^t ce^{A(t-s)}b\ell_{1k}(s)ds, \ldots, \int_0^t ce^{A(t-s)}b\ell_{Nk}(s)ds \right). \tag{6.22}$$

By setting

$$\hat{y} = (y_{1k}, \ldots, y_{Nk})^T, \tag{6.23}$$

$$C_k = [ce^{At_{1k}}x_0, \ldots, ce^{At_{Nk}}x_0]^T, \tag{6.24}$$

and

$$\hat{L}_k(y) = [L(t_{1k})(y), \ldots, L(t_{Nk})(y)]^T, \tag{6.25}$$

we can rewrite the equations as

$$(I + \lambda_k L(t))(y) + H_k\hat{y}$$
$$= \lambda_k L(t)(y_n) + H_k\hat{\alpha} + ce^{At}x_0\lambda_k\hat{L}_k(y) + (I + G_k)\hat{y}$$
$$= \lambda_k\hat{L}_k(y_n) + G_k\hat{\alpha} + [ce^{At_{1k}}x_0, \ldots, ce^{At_{Nk}}x_0]^T.$$

Since G_k is positive definite, so is $I + G_k$, and hence we can solve for $\hat{y}$ and develop the recursion

$$(y + \lambda_k L(t)(y - y_n)) + H_k\lambda_k(I + G_k)^{-1}\hat{L}_k(y_n - y)$$
$$= -H_k(I + G_k)^{-1}G_k\hat{\alpha} - (I + G_k)^{-1}C_k + H_k\hat{\alpha} + ce^{At}x_0.$$

Dividing by λ_k, we have

$$(\lambda_k^{-1}y_{n+1} + L(t)(y_{n+1} - y_n)) + H_k(I + G_k)^{-1}\hat{L}_k(y_n - y_{n+1})$$
$$= \lambda_k^{-1}(-H_k(I + G_k)^{-1}G_k\hat{\alpha} - (I + G_k)^{-1}C_k + H_k\hat{\alpha} + ce^{At}x_0).$$

The right-hand side is bounded and so, if the sequence of λ_k goes to infinity, then the right-hand side approaches 0. What remains to be proved is thus that the sequence of y_n is bounded.

6.6.2 The Discrete Case

In this case, the Hilbert space is

$$L_2[0, T] \times \mathbb{R}^{N+j},$$

with the norm defined as

$$\|(u; x)\|^2 = \int_0^T u^2(t)dt + x^T Q x,$$

where

$$Q = \begin{pmatrix} Q_1 & 0 \\ 0 & I \end{pmatrix}$$

and

$$Q_1 = \mathrm{Diag}(\lambda_k \omega_{1j}, \ldots, \lambda_k \omega_{jj}).$$

Since there are two different time sequences, we define

$$\wp_{ij}(s) = \begin{cases} ce^{A(r_{ij}-s)}b, & r_{ij} - s \geq 0, \\ 0 & \text{otherwise.} \end{cases}$$

The linear variety is

$$V_{x_0} = \Big\{ (u; (\rho, \gamma)) \, \Big| \, \rho_{ij} = ce^{A r_{ij}} x_0 + \int_0^T \wp_{ij}(s)u(s)ds,$$

$$\gamma_{ik} = ce^{A t_{ik}} x_0 + \int_0^T \ell_{ij}(s)u(s)ds \Big\}.$$

We define for typographical convenience

$$\hat{y} = \left(\int_0^T \wp_{1j}(s)u(s)ds, \ldots, \int_0^T \wp_{jj}(s)u(s)ds \right)^T,$$

$$\hat{x} = \left(\int_0^T \ell_{1j}(s)u(s)ds, \ldots, \int_0^T \ell_{Nj}(s)u(s)ds \right)^T,$$

$$\hat{\ell} = (\ell_{1N}(s), \ldots, \ell_{kN})^T$$

and

$$\hat{\wp} = (\wp_{1j}(s), \ldots, \wp_{jj}(s))^T.$$

The computation of $V_0^\perp$ reduces to the equation

$$\int_0^T u(s)v(s)ds + z^T Q_1 \hat{y} + w^T \hat{x} = 0,$$

which reduces to

$$V_0^\perp = \{(v, (z, w)) \mid v(s) + z^T Q_1 \hat{\wp} + w^T \hat{\ell} = 0\}.$$

As in the continuous case, to find the intersection of V_{x_0} and $V_0^\perp + p$, we must solve the following system of equations:

$$u(s) = -(\rho^T - \hat{y}^T)Q_1\hat{\wp} - (\gamma^T - \hat{\alpha}^T)\hat{\ell},$$

$$\rho_{ij} = ce^{Ar_{ij}}x_0 + \int_0^T \wp_{ij}(s)u(s)ds,$$

$$\gamma_{ik} = ce^{At_{ik}}x_0 + \int_0^T \ell_{ij}(s)u(s)ds.$$

Abbreviating notation, we have

$$u(s) = -(\rho^T - \hat{y}^T)Q_1\hat{\wp} - (\gamma^T - \hat{\alpha}^T)\hat{\ell},$$

$$\rho = C_r + \int_0^T \hat{\wp}u(s)ds,$$

$$\gamma = C_t + \int_0^T \hat{\ell}u(s)ds,$$

and substituting gives

$$\rho = C_r - \int_0^T \wp\wp^T Q_1\rho ds + \int_0^T \wp\wp^T Q_1\hat{y}ds - \int_0^T \wp\hat{\ell}^T \gamma ds$$
$$+ \int_0^T \wp\hat{\ell}^T \hat{\alpha}ds,$$

$$\gamma = C_t - \int_0^T \hat{\ell}\wp^T Q_1\rho ds + \int_0^T \hat{\ell}\wp^T Q_1\hat{y}ds - \int_0^T \ell\ell^T \gamma ds$$
$$+ \int_0^T \ell\ell^T \hat{\alpha}ds,$$

$$Q_1^{-1}Q_1\rho = C_r - \int_0^T \wp\wp^T Q_1\rho ds + \int_0^T \wp\wp^T Q_1\hat{y}ds - \int_0^T \wp\hat{\ell}^T \gamma ds$$
$$+ \int_0^T \wp\hat{\ell}^T \hat{\alpha}ds,$$

$$\gamma = C_t - \int_0^T \hat{\ell}\wp^T Q_1\rho ds + \int_0^T \hat{\ell}\wp^T Q_1\hat{y}ds - \int_0^T \ell\ell^T \gamma ds$$

$$+ \int_0^T \ell\ell^T \hat{a} ds,$$

$$Q_1^{-1} Q_1 \rho = C_r - \int_0^T \wp\wp^T ds Q_1 \rho + \int_0^T \wp\wp^T ds Q_1 \hat{y} - \int_0^T \wp\hat{\ell}^T ds\gamma$$
$$+ \int_0^T \wp\hat{\ell}^T ds\hat{a},$$

$$\gamma = C_t - \int_0^T \hat{\ell}\wp^T ds Q_1 \rho + \int_0^T \hat{\ell}\wp^T ds Q_1 \hat{y} - \int_0^T \ell\ell^T ds\gamma$$
$$+ \int_0^T \ell\ell^T ds\hat{a},$$

$$\begin{pmatrix} Q_1^{-1} + G & H \\ H^T & I+S \end{pmatrix} \begin{pmatrix} Q_1\rho \\ \gamma \end{pmatrix} = \begin{pmatrix} G & H \\ H^T & S \end{pmatrix} \begin{pmatrix} Q_1\hat{y}_n \\ \hat{a}_n \end{pmatrix} + \begin{pmatrix} C_r \\ C_t \end{pmatrix}. \tag{6.26}$$

We will now eliminate γ:

$$\begin{pmatrix} Q_1^{-1} + G - H(I+S)^{-1}H^T & 0 \\ (I+S)^{-1}H^T & I \end{pmatrix} \begin{pmatrix} Q_1\rho \\ \gamma \end{pmatrix}$$
$$= \begin{pmatrix} G - H(I+S)^{-1}H^T & H - H(I+S)^{-1}S \\ (I+S)^{-1}H^T & (I+S)^{-1}S \end{pmatrix} \begin{pmatrix} Q_1\hat{y}_n \\ \hat{a}_n \end{pmatrix}$$
$$+ \begin{pmatrix} C_r - H(I+S)^{-1}C_t \\ (I+S)^{-1}C_t \end{pmatrix}.$$

We thus arrive at

$$(I + GQ_1 - H(I+S)^{-1}H^T Q_1)\rho$$
$$= (GQ_1 - H(I+S)^{-1}H^T Q_1)\hat{y}_n$$
$$+ H(I+S)^{-1}\hat{a}_n + C_r - H(I+S)^{-1}C_t.$$

Going back to the original notation we have

$$(I + \lambda_n(GQ - H_n(I+S_n)^{-1}H_n^T Q))\hat{y}_{n+1}$$
$$= \lambda_n(GQ - H_n(I+S_n)^{-1}H_n^T Q)\hat{y}_n$$
$$+ H_n(I+S_n)^{-1}\hat{a}_n + C_r - H_n(I+S_n)^{-1}C_t,$$

which we rewrite as

$$\hat{y}_{n+1} = (\lambda_n^{-1} - A_n)^{-1} A_n \hat{y}_n + (\lambda_n^{-1} - A_n)^{-1}\lambda_n^{-1} U_n. \tag{6.27}$$

Rewriting this gives a less cumbersome form:

$$\hat{y}_{n+1} = (\epsilon_n - A_n)^{-1} A_n \hat{y}_n + (\epsilon_n - A_n)^{-1}\epsilon_n U_n. \tag{6.28}$$

We first consider the matrices A_n and U_n. These matrices come from a compact set since they just depend on the inner products of the ℓ_i and the $\wp_i$. The behavior of the matrices $(\epsilon_n - A_n)^{-1} A_n$ and $(\epsilon_n - A_n)^{-1} \epsilon_n$ reduce to eigenvalue calculations. We assume that we have chosen the parameters ϵ_n so that they approach zero.

Assumption 6.5

$$\{r_{ij} \mid i = 1, \ldots, j\} \cap \{t_{iN} \mid i = 1, \ldots, N\} = \emptyset$$

We now state and prove a technical lemma.

Lemma 6.6 *For every j the matrix*

$$G - H_j(I + S_j)^{-1} H_j^T$$

is nonsingular.

Proof. The matrix

$$\begin{pmatrix} G_j & H_j \\ H_j^T & S_j \end{pmatrix}$$

is the Grammian of a linearly independent set of vectors, by virtue of Assumption 6.5, and hence the matrix

$$\begin{pmatrix} G_j & H_j \\ H_j^T & I + S_j \end{pmatrix}$$

is positive definite.

Using elementary row and column operations we can reduce the matrix to

$$\begin{pmatrix} G_j - H_j(I + S_j)^{-1} H_j^T & 0 \\ (I + S_j)^{-1} H_j^T & I \end{pmatrix},$$

and the lemma follows.   ∎

The matrix A_j is likewise nonsingular for all j and so, by diagonalizing the A_j (or rather $A_j Q^{-1}$), we reduce the recursion to the form

$$z_{j+1} = (1 + \tau_j) z_j + \tau_j u_j, \tag{6.29}$$

where u_j is a bounded function of j.

Our goal is to show that we can choose the sequence of τ so that the solution of this difference equation is given by the infinite sum

$$\sum_{i=0}^{\infty} \tau_i,$$

which, moreover, is bounded.

The solution to (6.29) is of the form

$$z_j = \prod_{i=0}^{j}(1 + \tau_j) + \sum_{i=0}^{j} \prod_{k=i+1}^{j} (1 + \tau_k)u_i,$$

so in the limit the following terms must exist:

$$\prod_{i=0}^{\infty}(1 + \tau_i) \tag{6.30}$$

and

$$\sum_{i=0}^{\infty} \prod_{k=i+1}^{\infty} (1 + \tau_k)\tau_i u_i. \tag{6.31}$$

It is well known that product of (6.30) is finite provided that

$$\sum_{i=0}^{\infty} \tau_i$$

converges (see [83]). The conditions for the sum of (6.31) to converge are a little more obscure. It is easy to prove that if the ratio test proves the convergence of

$$\sum_{i=0}^{\infty} \tau_i,$$

then the series in (6.31) exists and is finite.

SUMMARY

In this book we present a general approach to interpolating and smoothing splines that includes all polynomial, exponential, and trigonometric splines. In this chapter, we proved that with system dynamics governed by ordinary differential equations, two-point boundary value problems, or multipoint boundary value problems, the theory is essentially the same. We have shown that the solution to all of these problems reduces to finding a point on a linear variety that is nearest in a suitable Hilbert space norm to a given data point. In this chapter we have worked with continuous data, which complicates implementation, but the problem does arise in applications. The approach would in principle work in a general Banach space, but in practice the equations are not solvable. However, the answers obtained by restricting to a Hilbert space are satisfactory for most problems of interest.

Chapter Seven

MONOTONE SMOOTHING SPLINES

In this chapter, we consider a variation to the basic theme of this book, by imposing certain types of regularity conditions on the produced curves that have to hold for all times–namely, monotonicity conditions. As such, the solution to the problem of generating curves by driving the output of a particular nilpotent single-input, single-output linear control system close to given waypoints is analyzed, when the curves are constrained by an infinite-dimensional nonnegativity constraint on one of the derivatives of the curve. The main theorem in this chapter states that the optimal curve is a piecewise polynomial of known degree. For the two-dimensional case, this problem is completely solved when the acceleration is controlled directly. The solution is obtained by exploiting a finite reparameterization of the problem that can be solved using dynamic programming.

In many cases, the type of construction we have seen so far in this book is not enough since one sometimes wants the generated curve to exhibit certain structural properties, such as monotonicity or convexity (see [17],[21],[41], [47],[49],[66],[76],[85]). These properties correspond to nonnegativity constraints on the first and second derivatives of the curve, respectively, and hence the nonnegative derivative constraint will be the main focus of this chapter.

7.1 THE MONOTONE SMOOTHING PROBLEM

Consider the problem of constructing a curve that passes close to given data points, at the same time that we want the curve to exhibit certain monotonicity properties. In other words, if $p(t)$ is our curve, we want $(p(t_i)-\alpha_i)^2$, $i = 1, \ldots, m$, to be qualitatively small. Here, $(t_1, \alpha_1), \ldots, (t_m, \alpha_m)$ are the data points, with $\alpha_i \in \mathbb{R}$, $i = 1, \ldots, m$, and $0 < t_1 < t_2 < \cdots < t_m \leq T$, for some given terminal time $T > 0$. We do not only, however, want to keep the interpolation errors small. We also want the curve to vary in a smooth way, and that

$$p^{(n)}(t) \geq 0, \ \forall t \in [0, T], \tag{7.1}$$

for some given positive integer n. We now revisit the conditions for control theoretic smoothing splines.

Let

$$A = \begin{pmatrix} 0 & 1 & 0 & \cdots & 0 \\ 0 & 0 & 1 & \cdots & 0 \\ & \vdots & & \ddots & \vdots \\ 0 & 0 & 0 & \cdots & 1 \\ 0 & 0 & 0 & \cdots & 0 \end{pmatrix}, \quad b = \begin{pmatrix} 0 \\ \vdots \\ 0 \\ 1 \end{pmatrix}, \quad (7.2)$$

and

$$\begin{aligned} c_1 &= (\; 1 \quad 0 \quad \cdots \quad 0 \;), \\ c_2 &= (\; 0 \quad 0 \quad \cdots \quad 1 \;), \end{aligned} \quad (7.3)$$

where A is an $n \times n$ matrix, b is $n \times 1$, and c_1 and c_2 are $1 \times n$. Here, $c_1 x(t)$ takes on the role of $p(t)$, and by our particular choices of A and b in (7.3), x is a vector of successive derivatives.

The problem can now be cast as

$$\inf_u \left\{ \frac{1}{2} \int_0^T u^2(t)dt + \frac{1}{2} \sum_{i=1}^m w_i (c_1 x(t_i) - \alpha_i)^2 \right\}, \quad (7.4)$$

subject to

$$\begin{cases} \dot{x} = Ax + bu, \; x(0) = 0, \\ u \in L_2[0, T], \\ c_2 x(t) \geq 0, \; \forall t \in [0, T], \end{cases} \quad (7.5)$$

where $w_i \geq 0$ reflects how important it is that the curve passes close to a particular $\alpha_i \in \mathbb{R}$. The reason for infimizing rather than minimizing is that it is not clear, at this point, that the minimizer exists.

Now, since $\dot{x} = Ax + bu$, $c_1 x(t_i)$ is given by

$$c_1 x(t_i) = \int_0^{t_i} c_1 e^{A(t_i - t)} bu(t)dt,$$

since $x(0) = 0$. This expression can furthermore be written as

$$c_1 x(t_i) = \int_0^T \ell_i(t)u(t)dt,$$

where, as before, we make use of the linearly independent basis functions

$$\ell_i(t) = \begin{cases} c_1 e^{A(t_i - t)} b & \text{if } t \leq t_i, \\ 0 & \text{if } t > t_i, \end{cases} \quad i = 1, \dots, m. \quad (7.6)$$

Our infimization over u can then be rewritten as

$$\inf_u \left\{ \frac{1}{2} \int_0^T u^2(t)dt + \frac{1}{2} \sum_{i=1}^m w_i \left(\int_0^T \ell_i(t)u(t)dt - \alpha_i \right)^2 \right\}, \qquad (7.7)$$

which is an expression that depends only on u.

Since we want $c_2 x(t)$ to be continuous, we let the constraint space be $C[0, T]$, that is, the space of continuous functions. In a similar fashion as before, we can express $c_2 x(t)$ as

$$c_2 x(t) = \int_0^t c_2 e^{A(t-s)} bu(s)ds = \int_0^t f(t, s)u(s)ds.$$

This allows us to form the associated Lagrangian [64]

$$L(u, \nu) = \frac{1}{2} \int_0^T u^2(t)dt + \frac{1}{2} \sum_{i=1}^m w_i \left(\int_0^T \ell_i(t)u(t)dt - \alpha_i \right)^2$$
$$- \int_0^T \int_0^t f(t, s)u(s)ds d\nu(t), \qquad (7.8)$$

where $\nu \in BV[0, T]$ (the space of functions of bounded variations, which is the *dual space* of $C[0, T]$). The optimal solution to our original optimization problem is thus found by solving

$$\max_{0 \leq \nu \in BV[0,T]} \inf_{u \in L_2[0,T]} L(u, \nu). \qquad (7.9)$$

7.2 PROPERTIES OF THE SOLUTION

Lemma 7.1 *Let $(\tilde{A}, \tilde{b}, \tilde{c})$ be a triple, where $\tilde{A}$ is an $n \times n$ matrix, $\tilde{b}$ is $n \times 1$, and $\tilde{c}$ is $1 \times n$. If $\dot{x} = \tilde{A}x + \tilde{b}u$, $x(0) = 0$, then the set of controls in $L_2[0, T]$ that make the solution to the differential equation satisfy*

$$\tilde{c}x(t) \geq 0, \ \forall t \in [0, T]$$

is a closed, nonempty, and convex set.

Proof. We first show convexity. Given two $u_i(t) \in L_2[0, T]$, $i = 1, 2$, such that

$$\int_0^t \tilde{c}e^{\tilde{A}(t-s)}\tilde{b}u_i(s)ds \geq 0, \ \forall t \in [0, T], \ i = 1, 2,$$

then for any $\lambda \in [0, 1]$ we have

$$\int_0^t \tilde{c}e^{\tilde{A}(t-s)}\tilde{b}(\lambda u_1(s) + (1 - \lambda)u_2(s))ds \geq 0, \ \forall t \in [0, T],$$

and convexity thus follows.

Now, consider a collection of controls $\{u_i(t)\}_{i=0}^{\infty}$, where each individual control makes the solution to the differential equation satisfy $\tilde{c}x(t) \geq 0 \, \forall t \in [0, T]$, and where $u_i \rightarrow \hat{u}$ as $i \rightarrow \infty$. But, due to the compactness of $[0, t]$, we have that

$$\lim_{i \to \infty} \int_0^t \tilde{c}e^{\tilde{A}(t-s)}\tilde{b}u_i(s)ds = \int_0^t \tilde{c}e^{\tilde{A}(t-s)}\tilde{b}\hat{u}(s)ds \geq 0, \quad \forall t \in [0, T].$$

The fact that $L_2[0, T]$, with the natural norm defined on it, is a Banach space gives us that the limit, $\hat{u}$, still remains in that space. The set of admissible controls is thus closed.

Furthermore, since $x(0) = 0$, we can always let $u \equiv 0$. This gives that the set of admissible controls is nonempty, which concludes the proof. ∎

Lemma 7.2 *The cost functional in (7.4) is convex in u.*

The proof of this lemma is trivial since both terms in (7.4) are quadratic functions of u.

Lemmas 7.1 and 7.2 are desirable in any optimization problem since they are strong enough to guarantee the existence of a unique optimal solution (see, e.g., [64]), and we can thus replace inf in (7.8) with min, which directly allows us to state the following standard theorem about the optimal control.

Theorem 7.3 *There is a unique $u_0 \in L_2[0, T]$ that solves the optimal control problem in (7.4).*

We omit the proof of this and refer to any textbook on optimization theory for the details. (See for example [64].)

Lemma 7.4 *Given the optimal solution u_0, the optimal $\nu_0 \in BV[0, T]$, $\nu_0 \geq 0$, varies only where $c_2 x(t) = 0$. On intervals where $c_2 x(t) > 0$, $\nu_0(T) - \nu_0(t)$ is a nonnegative, real constant.*

Proof. Since $\nu_0(T) - \nu_0(t) \geq 0$ due to the positivity constraint on ν_0, we reduce the value of the Lagrangian in (7.8) whenever ν_0 changes, except when $c_2 x(t) = 0$. But, since ν_0 maximizes $L(u_0, \nu)$, we only allow ν_0 to change when $c_2 x(t) = 0$, and the lemma follows. ∎

Now, before we can completely characterize the optimal control solution, one observation to be made is that

$$c_2 x(t) = \begin{pmatrix} 0 & 0 & \cdots & 1 \end{pmatrix} x(t) = \int_0^t u(s)ds,$$

that is, $f(t, s)$ is in fact equal to 1 in (7.8). This allows us to rewrite the Lagrangian as

$$L(u, \nu) = \frac{1}{2} \int_0^T u^2(t)dt + \frac{1}{2} \sum_{i=1}^m \omega_i \left(\int_0^T \ell_i(t)u(t)dt - \alpha_i \right)^2$$
$$- \int_0^T \int_0^t u(s)dsd\nu(t). \tag{7.10}$$

By integrating the Stieltjes integral in (7.10) by parts, we can furthermore reduce the Lagrangian to

$$L(u, \nu) = \frac{1}{2} \int_0^T u^2(t)dt + \frac{1}{2} \sum_{i=1}^m \omega_i \left(\int_0^T \ell_i(t)u(t)dt - \alpha_i \right)^2$$
$$- \int_0^T (\nu(T) - \nu(t))u(t)dt, \tag{7.11}$$

which is a more easily manipulated expression.

Definition 7.5 *Let $PP^k[0, T]$ denote the set of piecewise polynomials of degree k on $[0, T]$. Let $P^k[0, T]$ denote the set of polynomials of degree k on that interval.*

Theorem 7.6 *The control in $L_2[0, T]$ that minimizes the cost in (7.4) is in $PP^n[0, T]$. It furthermore changes from different polynomials of degree n only at the interpolation times, t_i, $i = 1, \ldots, m$, and at times when $c_2 x(t)$ changes from $c_2 x(t) > 0$ to $c_2 x(t) = 0$ and vice versa.*

Proof. Due to the convexity of the problem and the existence and uniqueness of the solution, we can obtain the optimal controller by calculating the Gateaux derivative of L with respect to u and setting this equal to zero for all increments $h \in L_2[0, T]$.

By letting $L_\nu(u) = L(u, \nu)$, we get that

$$\delta L_\nu(u, h) = \lim_{\epsilon \to 0} \frac{1}{\epsilon}(L_\nu(u + \epsilon h) - L_\nu(u))$$

is given by the integral

$$\int_0^T \left(u(t) + \sum_{i=1}^m \omega_i \left(\int_0^T \ell_i(s)u(s)ds - \alpha_i \right) \ell_i(t) - (\nu(T) - \nu(t)) \right) h(t)dt. \tag{7.12}$$

For the expression in (7.12) to be zero for all $h \in L_2[0, T]$, we need to have that

$$u_0(t) + \sum_{i=1}^{m} \omega_i \left(\int_0^T \ell_i(s) u_0(s) ds - \alpha_i \right) \ell_i(t) - (\nu(T) - \nu(t)) = 0.$$

This especially has to be true for $\nu = \nu_0$, which gives that

$$u_0(t) + \sum_{i=1}^{m} \omega_i \left(\int_0^T \ell_i(s) u_0(s) ds - \alpha_i \right) \ell_i(t) - C_j = 0, \qquad (7.13)$$

whenever $c_2 x_0(t) > 0$. Here C_j is a constant. The index j indicates that this constant differs on different intervals where $c_2 x_0(t) > 0$.

Now, the integral terms in (7.13) do not depend on t, while $\ell_i(t)$ is in $\mathcal{P}^n[0, t_i]$ for $i = 1, \ldots, m$. This, combined with the fact that $\nu_0(T) - \nu_0(t) = C_j$ if $\dot{x}(t) > 0$, directly gives us that the optimal control, $u_0(t)$, has to be in $P\mathcal{P}^n[0, T]$. It obviously changes at the interpolation times, due to the shape of the ℓ_i, but it also changes if C_j changes, that is, it changes if $c_2 x_0(t) = 0$. It should be noted that if $c_2 x_0(t) \equiv 0$ on an interval, $\nu_0(t)$ may change on the entire interval, but since $c_2 x_0(t) \equiv 0$ we also have that $u_0(t) \equiv 0$ on the interior of this interval. But a zero function is, of course, polynomial. Thus we know that our optimal control is in $P\mathcal{P}^n[0, t]$, and the theorem follows.   ∎

Corollary 7.7 *If $n = 2$, then the optimal control is piecewise linear (in $P\mathcal{P}^1[0, T]$), with changes from different polynomials of degree one at the interpolation times, and at times when $c_2 x(t)$ changes from $c_2 x(t) > 0$ to $c_2 x(t) = 0$ and vice versa.*

7.3 DYNAMIC PROGRAMMING

Based on the general properties of the solution, the idea now is to formulate the monotone interpolation problem as a finite-dimensional programming problem that can be dealt with efficiently. If we drive the system $\dot{x} = Ax + bu$, where A and b are defined in (7.3), between x_i and x_{i+1} on the time interval $[t_i, t_{i+1}]$, under the constraint $c_2 x(t) \geq 0$, we see that we must at least have

$$\begin{aligned} c_2 x_i &\geq 0, \\ c_2 x_{i+1} &\geq 0, \\ D(x_{i+1} - x_i) &\geq 0, \end{aligned} \qquad (7.14)$$

where

$$D = \begin{pmatrix} 1 & 0 & \cdots & 0 & 0 \\ 0 & 1 & \cdots & 0 & 0 \\ \vdots & & \ddots & & \vdots \\ 0 & 0 & \cdots & 1 & 0 \end{pmatrix},$$

and the inequality in (7.14) is taken component-wise. We denote the constraints in (7.14) by

$$\mathcal{D}(x_i, x_{i+1}) \geq 0.$$

Since the original cost functional in (7.4) can be divided into one interpolation part and one smoothing part, it seems natural to define the following *optimal value function* as

$$\begin{cases} \hat{S}_i(x_i) \\ \quad = \min_{x_{i+1} | \mathcal{D}(x_i, x_{i+1}) \geq 0} \left\{ V_i(x_i, x_{i+1}) + \hat{S}_{i+1}(x_{i+1}) \right\} + \omega_i (c_1 x_i - \alpha_i)^2, \\ \\ \hat{S}_m(x_m) = \omega_m (c_1 x_m - \alpha_m)^2, \end{cases}$$

(7.15)

where $V_i(x_i, x_{i+1})$ is the cost for driving the system between x_i and x_{i+1} using a control in $P\mathcal{P}^n[t_i, t_{i+1}]$, while keeping $c_2 x(t)$ nonnegative on the time interval $[t_i, t_{i+1}]$.

The optimal control problem thus becomes that of finding $\hat{S}_0(0)$, where we let $\omega_0 = 0$, while α_0 can be any arbitrary number. In light of Theorem 7.6, this problem is equivalent to the original problem, and if $V_i(x_i, x_{i+1})$ could be uniquely determined, it would correspond to finding the $n \times m$ variables $x_1, \ldots, x_m$, which is a *finite-dimensional* reparameterization of the original, infinite-dimensional programming problem.

For this dynamic programming approach to work, our next task becomes that of determining the function $V_i(x_i, x_{i+1})$. Even though that is typically not an easy problem, a software package for computing approximations of such monotone polynomials was developed in [17]. In [47],[49] this problem of exact interpolation, over piecewise polynomials, of convex or monotone data points was furthermore investigated from a theoretical point of view. It is thus our belief that showing that the original problem can formulated as a dynamic programming problem involving exact interpolation is a valuable result since it greatly simplifies the structure of the problem. It furthermore transforms it to a form that has been extensively studied in the literature.

7.4 MONOTONE CUBIC SPLINES

If we change our notation slightly in such a way that our state variable is given by $(x, \dot{x})$, $x, \dot{x} \in \mathbb{R}$, the dynamics of the system becomes

$$\ddot{x} = u.$$

The optimal value function in (7.15) thus takes on the form

$$
\begin{aligned}
\hat{S}_i(x_i, \dot{x}_i) & \\
&= \min_{\substack{x_{i+1} \geq x_i \\ \dot{x}_{i+1} \geq 0}} \left\{ V_i(x_i, \dot{x}_i, x_{i+1}, \dot{x}_{i+1}) + \hat{S}_{i+1}(x_{i+1}, \dot{x}_{i+1}) \right\} + w_i(x_i - \alpha_i)^2 \\
\hat{S}_m(x_m, \dot{x}_m) &= w_m(x_m - \alpha_m)^2.
\end{aligned}
$$

$$(7.16)$$

7.4.1 Two-Point Interpolation

Given the times t_i and t_{i+1}, the positions x_i and x_{i+1}, and the corresponding derivatives $\dot{x}_i$ and $\dot{x}_{i+1}$, the question to be answered, as indicated by Corollary 7.7, is the following. *How do we drive the system between $(x_i, \dot{x}_i)$ and $(x_{i+1}, \dot{x}_{i+1})$, with a piecewise linear control input that changes between different polynomials of degree one, only when $\dot{x}(t) = 0$, in such a way that $\dot{x}(t) \geq 0 \ \forall t \in [t_i, t_{i+1}]$, while minimizing the integral over the square of the control input?* Without loss of generality, for notational purposes, we translate the system and rename the variables so that we want to produce a curve defined on the time interval $[0, t_F]$ between $(0, \dot{x}_0)$ and $(x_F, \dot{x}_F)$.

Assumption 7.8

$$\dot{x}_0, \dot{x}_F \geq 0, \ x_F > 0, \ t_F > 0.$$

It should be noted that if $x_F = 0$, and either $\dot{x}_0 > 0$ or $\dot{x}_F > 0$, then $\dot{x}(t)$ can never be continuous. This case has to be excluded since we demand that our constraint space be $C[0, T]$. If, furthermore, $x_F = \dot{x}_0 = \dot{x}_F = 0$, then the optimal control is obviously given by $u \equiv 0$ on the entire interval.

One first observation is that the optimal solution to this *two-point interpolation problem* is to use standard cubic splines if that is possible, that is, if $\dot{x}(t) \geq 0$ for all $t \in [0, t_F]$. In this well-studied case (see [6],[85]), we simply have that

$$x(t) = \frac{1}{6}at^3 + \frac{1}{2}bt^2 + \dot{x}_0 t, \qquad (7.17)$$

where

$$\begin{pmatrix} a \\ b \end{pmatrix} = \frac{6}{t_F^3} \begin{pmatrix} t_F(\dot{x}_0 + \dot{x}_F) - 2x_F \\ t_F x_F - \frac{1}{3}t_F^2(2\dot{x}_0 + \dot{x}_F) \end{pmatrix}. \tag{7.18}$$

This solution corresponds to having $\nu(t) = \nu(t_{i+1})$, for all $t \in [t_i, t_{i+1})$ in (7.11), and it gives the total cost

$$\mathcal{I}_1 = \int_0^{t_F} (at + b)^2 dt = 4\frac{(\dot{x}_0 t_F^2 - 3x_F t_F)(\dot{x}_0 + \dot{x}_F) + 3x_F^2 + t_F^2 \dot{x}_F^2}{t_F^3},$$

$$\tag{7.19}$$

where the subscript 1 denotes the fact that only one polynomial of degree one was used to compose the second derivative.

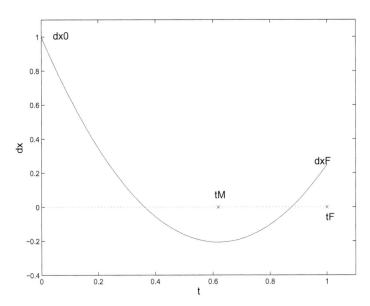

Figure 7.1 The case where a cubic spline cannot be used if the derivative has to be nonnegative. Plotted is the derivative that clearly intersects $\dot{x} = 0$.

However, not all curves can be produced by such a cubic spline if the curve has to be nondecreasing at all times. Given Assumption 7.8, the one case where we can not use a cubic spline can be seen in Figure 7.1, and from geometric considerations we get four different conditions that all need to hold for the derivative to be negative. These necessary and sufficient conditions are

$$\begin{array}{ll} \text{(i)} & a > 0, \\ \text{(ii)} & b < 0, \\ \text{(iii)} & \ddot{x}(t_M) < 0, \\ \text{(iv)} & t_M < t_F, \end{array} \tag{7.20}$$

where a and b are defined in (7.17), and t_M is defined in Figure 7.1.

We can now state the following lemma.

Lemma 7.9 *Given Assumption 7.8, a standard cubic spline can be used to produce monotonously increasing curves if and only if*

$$x_F \geq \chi(t_F, \dot{x}_0, \dot{x}_F) = \frac{t_F}{3}(\dot{x}_0 + \dot{x}_F - \sqrt{\dot{x}_0\dot{x}_F}). \tag{7.21}$$

The proof of this follows from simple algebraic manipulations [34], and we now need to investigate what the optimal curve looks like in the case when we cannot use standard cubic splines.

7.4.2 Monotone Interpolation

Given two points such that $x_F < \chi(t_F, \dot{x}_0, \dot{x}_F)$, how should the interpolating curve be constructed so that the second derivative is piecewise linear, with switches only when $\dot{x}(t) = 0$? One first observation is that it is always possible to construct a piecewise polynomial path that consists of three polynomials of degree one and respects the interpolation constraint, and in what follows we will see that such a path also respects the monotonicity constraint.

The three interpolating polynomials are given by

$$u(t) = \begin{cases} a_1 t + b_1 & \text{if } 0 \leq t < t_1, \\ 0 & \text{if } t_1 \leq t < t_2, \\ a_2(t - t_2) + b_2 & \text{if } t_2 \leq t \leq t_F, \end{cases} \tag{7.22}$$

where

$$\begin{pmatrix} a_1 \\ b_1 \end{pmatrix} = \frac{6}{t_1^3}\begin{pmatrix} t_1\dot{x}_0 - 2x_1 \\ t_1 x_1 - 2/3 t_1^2 \dot{x}_0 \end{pmatrix},$$

$$\begin{pmatrix} a_2 \\ b_2 \end{pmatrix} = \frac{6}{(t_F-t_2)^3}\begin{pmatrix} (t_F - t_2)\dot{x}_F - 2(x_F - x_1) \\ (t_F - t_2)(x_F - x_1) - 1/3(t_F - t_1)^2\dot{x}_F \end{pmatrix}, \tag{7.23}$$

and where $x(t_1) = x(t_2) = x_1$, together with t_1 and t_2, is a parameter that needs to be determined.

Assumption 7.10

$$\dot{x}_0, \dot{x}_F, x_F, t_F > 0.$$

We need this assumption, which is stronger than Assumption 7.8, for the following lemma, but it should be noted that if $\dot{x}_0 = 0$ or $\dot{x}_F = 0$ we would then just let the first or the third polynomial on the curve be zero.

We now state the possibility of such a feasible three-polynomial construction.

Lemma 7.11 *Given* $(t_F, \dot{x}_0, x_F, \dot{x}_F)$ *such that* $x_F < \chi(t_F, \dot{x}_0, \dot{x}_F)$, *then a feasible, monotone curve will be given by (7.22) as long as Assumption 7.10 holds. Furthermore, the optimal* $t_1, t_2,$ *and* x_1 *are given by*

$$\begin{cases} t_1 = 3x_1/\dot{x}_0, \\ t_2 = t_F - 3(x_F - x_1)/\dot{x}_F, \\ x_1 = \dot{x}_0^{3/2} x_F/(\dot{x}_0^{3/2} + \dot{x}_F^{3/2}). \end{cases} \tag{7.24}$$

The proof is constructive and is based on showing that, with the type of construction given in (7.22), the optimal choice of t_1, t_2, x_1 gives a feasible curve. We refer the reader to [34] for the details. We can thus construct a feasible path, as seen in Figure 7.2, by using three polynomials whose second derivatives are linear.

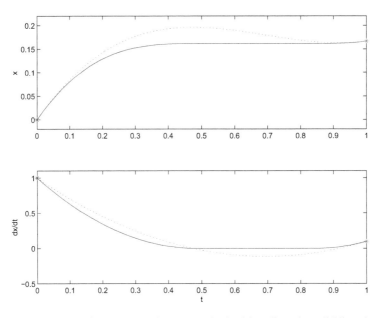

Figure 7.2   The dotted line corresponds to a standard cubic spline; the solid line shows the three construction from Lemma 7.11. Depicted are the position and the velocity.

Theorem 7.12 (Monotone Interpolation) *Given Assumption 7.8, the optimal control that drives the path between* $(0, \dot{x}_0)$ *and* $(x_F, \dot{x}_F)$ *is given by (7.17) if* $x_F \geq \chi(t_F, \dot{x}_0, \dot{x}_F)$ *and by (7.22) otherwise.*

Proof. The first part of the theorem is obviously true. If we can construct a standard, cubic spline, then this is optimal. However, what we need to show is that when $x_F < \chi(t_F, \dot{x}_0, \dot{x}_F)$ the path given in (7.22) is in fact optimal.

The cost for using a path given in (7.22) is

$$\mathcal{I}_3 = \int_0^{t_1} (a_1 t + b_1)^2 dt + \int_{t_2}^{t_F} (a_2(t - t_2) + b_2)^2 dt = \frac{4(\dot{x}_F^{3/2} + \dot{x}_0^{3/2})^2}{9 x_F},$$

where the coefficients are given in (7.24). We now add another, arbitrary polynomial, as seen in Figure 7.3, to the path as

$$u(t) = \begin{cases} a_1 t + b_1 & \text{if } 0 \le t < t_1, \\ 0 & \text{if } t_1 \le t < t_3, \\ a_3(t - t_3) + b_3 & \text{if } t_3 \le t < t_4, \\ 0 & \text{if } t_4 \le t < t_2, \\ a_2(t - t_2) + b_2 & \text{if } t_2 \le t \le t_F, \end{cases} \tag{7.25}$$

where $0 < t_1 \le t_3 \le t_4 \le t_2 < t_F$. Furthermore, t_3, t_4, and $x_2 = x(t_4)$ (see Figure 7.3) are chosen arbitrarily, while the old variables, t_1, t_2, and $x_1 = x(t_1)$, are defined to be optimal with respect to the new, translated end-conditions that the extra polynomials give rise to.

After some straightforward calculations, we get that the cost for this new path is

$$\mathcal{I}_5 = \frac{4(\dot{x}_F^{3/2} + \dot{x}_0^{3/2})^2}{9(x_F - x_2)} + \frac{12(x_2 - x_1)^2}{(t_4 - t_3)^3}, \tag{7.26}$$

where the subscript 5 denotes the fact that we are now using five polynomials of degree one to compose our second derivative. It can be seen that we minimize $\mathcal{I}_5$ if we let $x_2 = x_1$ and make $t_4 - t_3$ as large as possible. This corresponds to letting $t_3 = t_1$ and $t_4 = t_2$, which gives us the old solution from Lemma 7.11, defined in (7.22). ∎

7.4.3 Monotone Cubic Smoothing Splines

We now have a way of producing the optimal, monotone path between two points, while controlling the acceleration directly. We are thus ready to formulate the transition cost function in (7.16), $V_i(x_i, \dot{x}_i, x_{i+1}, \dot{x}_{i+1})$, that defines the cost for driving the system between $(x_i, \dot{x}_i)$ and $(x_{i+1}, \dot{x}_{i+1})$, with minimum energy, while keeping the derivative nonnegative.

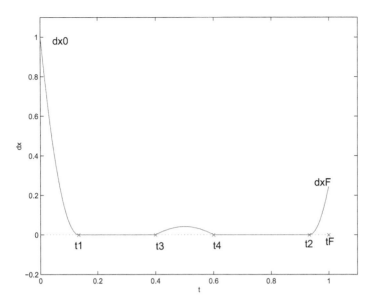

Figure 7.3 Two extra polynomials are added to the produced path. Depicted is the derivative of the curve.

Based on Theorem 7.12 and given Assumption 7.8, we have that[1]

$V_i(x_i, \dot{x}_i, x_{i+1}, \dot{x}_{i+1})$

$$
= \begin{cases}
4\dfrac{\dot{x}_i(t_{i+1}-t_i)^2 - 3(x_{i+1}-x_i)(t_{i+1}-t_i)(\dot{x}_i+\dot{x}_{i+1}) + 3(x_{i+1}-x_i)^2 + (t_{i+1}-t_i)^2\dot{x}_{i+1}^2}{(t_{i+1}-t_i)^3} \\[2mm]
\qquad \text{if } x_{i+1} - x_i \geq \chi(t_{i+1} - t_i, \dot{x}_i, \dot{x}_{i+1}), \\[4mm]
\dfrac{4(\dot{x}_{i+1}^{3/2} + \dot{x}_i^{3/2})^2}{9(x_{i+1}-x_i)} \\[2mm]
\qquad \text{if } x_{i+1} - x_i < \chi(t_{i+1} - t_i, \dot{x}_i, \dot{x}_{i+1}),
\end{cases}
$$

$$(7.27)$$

where $t_0 = x_0 = \dot{x}_0 = 0$.

If we use this cost in the dynamic programming algorithm formulated in (7.16), we get the results displayed in Figures 7.4–7.5, which shows that our approach does not only work in theory, but also in practice.

It should be noted that a major challenge is the construction of monotone smoothing splines of higher degree than cubic. For example, in some applications it is necessary to construct a smoothing spline such that $y(t) \geq 0$.

---

[1] If $x_{i+1} - x_i = \dot{x}_i = \dot{x}_{i+1} = 0$, then the optimal control is obviously zero, meaning that $V_i(x_i, \dot{x}_i, x_{i+1}, \dot{x}_{i+1}) = 0$.

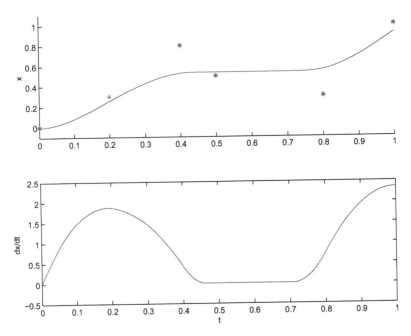

Figure 7.4 Monotone smoothing splines with $\omega_i = 1000, i = 1, \ldots, 5$.

This can be solved by approximating $z(t) = \int_0^t y(s)ds$ from the data. But, since the data are defined over y, some approximation scheme is needed to obtain the integral expression for z, which can be achieved with a monotone spline for approximating y by $\dot{z}$. If z is only cubic (as in the previous sections), y is only quadratic, in which case convergence is bound to be slow.

7.5 PROBABILITY DENSITIES

The monotonicity constraint can be thought of as a special type of continuous constraint (it has to hold for all times), and there are other cases in which such constraints must be handled. We here investigate that topic in the context of approximating probability densities from data.

There is ongoing research in the applications and theory of smoothing splines in statistics. Work by Peter Hall [45] and the book by R. Eubank [36] are examples of really good work in the area. Eubank's book is an excellent up-to-date place to become familiar with the work in the area. In this section, we propose to use control theoretic splines to attack a series of problems in the theory of density estimation, as well as outline some possible solution directions to a collection of fundamental problem.

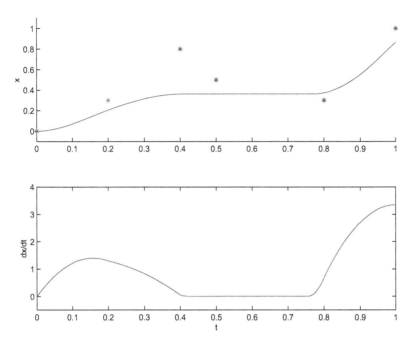

Figure 7.5 Monotone smoothing splines with $w_4 = 10w_i, i \neq 4$ (with $t_4 = 0.8$), resulting in a different curve from that in Figure 7.4, where equal importance is given to all of the waypoints.

Recalling the definition of the cubic smoothing spline (as in [96]), we have

$$J(\ddot{y}) = \lambda \int_0^T \ddot{y}^2(t)dt + \frac{1}{N}\sum_{i=1}^N (y(t_i) - \alpha_i)^2,$$

where the data set is

$$D = \{(t_i, \alpha_i) \mid i = 1, \ldots, N\}.$$

This cost function is minimized over $L_2[0, T]$ to obtain a curve $y(t)$ which is the smoothing spline. We have already seen that $y(t)$ is the classic interpolating cubic spline with a new set of data, where the new data $\hat{\alpha_i} = y(t_i)$ are determined by the minimization. In this formulation, there is one design parameter, λ. The smoothing spline procedure can be thought of as a least squares minimization based on a norm that is determined by the values t_i, and then using the estimated values to construct a cubic spline. This point of view is established and exploited in [102].

In this section, we consider two main sets of problems. The first is the estimation of probability distributions from sampled data using splines; the second is the estimation of exponential distributions using splines to estimate the log distribution. The work of Kooperberg and Stone [46],[56] is directly related to the second set of problems. Also the papers by Hall and co-authors [13] and Mammen [65] are directly relevant.

7.5.1 Estimation of Probability Distributions

This is, of course, an area of statistics in which there is a tremendous amount of literature. Surprisingly, there are two distinct bodies of literature, the statistics literature and the systems identification literature of control theory, and the two bodies of work only occasionally meet.

> **Problem 1**: Given the empirical distribution, use splines to construct a smooth version of the distribution.

We are not concerned with the specific construction of the empirical distribution at this point. We simply assume that it is given and that there are a limited number of levels. A rule of thumb in the use of splines is that no more than 10 nodal points should be used. This is because there is considerable linear algebra involved and many matrix methods start to show instability at around dimension 10. Many of the algorithms are required to solve systems with dimension twice the number of nodes.

Given the empirical distribution, we choose the midpoint of each interval and the value of the distribution at that point as the datum (t_i, α_i). Here we assume that the probability distribution function has finite support on $[0, T]$. Without loss of generality, we assume that $0 \leq t_i \leq T$ and we use the cost function

$$J(u, x_0) = \lambda \int_0^T u^2(t)dt + \sum_{i=1}^N w_i(y(t_i) - \alpha_i)^2 + x_0' Q x_0.$$

We impose the added constraint that

$$\int_0^T y(t)dt = 1.$$

Our constraint variety is then

$$V_c = \left\{ (u; x_0; \hat{y}) \,\middle|\, y(t_i) = ce^{At_i} x_0 + \int_0^{t_i} ce^{A(t_i-s)} u(s)ds, \right.$$
$$\left. \int_0^T [ce^{At} x_0 + \int_0^t ce^{A(t-s)} bu(s)ds]dt \right\}.$$

The construction of the spline is a straightforward application of the basic algorithm. We also impose a post hoc constraint that

$$y(t) \geq 0.$$

In the above construction, there is nothing to prevent the spline from being negative on intervals. This can be remedied by ad hoc methods of looking at the spline and, at the points where it becomes negative, imposing the point constraint that the derivative at that point must be positive. This is the usual technique. A more global method can be applied by insisting that the inequality

$$y(t) \geq 0$$

be satisfied on the domain. In fact, in the previous chapter (as well as in [31]), we saw how to enforce such monotonicity constraints. (The downside of the method in Chapter 7 is that the problem changes from a minimum norm problem in Hilbert space to a dynamic programming problem.)

An alternative method is to use the fact that the control theoretic splines are an approximation of a linear filter of the form

$$y(t) = \int_0^t k(t, s) f(s) ds.$$

This filter is developed in detail in the previous sections. Using this filter, it is not necessary to impose restrictions on the empirical distribution. The integral is to be evaluated using any good quadrature scheme, and the integration can thus be as accurate as we like.

Problem 2: Choose the form of the empirical distribution.

Here we assume that we have sampled the distribution a large number of times. We know that the choice of intervals makes a significant difference in the form of the empirical distribution. Gross features are usually relatively clear, but there can be small differences that are lost if too few subintervals are chosen. Smoothing splines tend to wash out small differences unless the weights are chosen carefully. The choice of λ also makes a very large difference in the shape of the smoothing spline.

In [53], the parameter was chosen in several different ways to reflect long-term vs. short-term effects in the Dow Jones Industrial Average over a 20-year period. Choosing λ to be large forces the control to be small and allows for errors in the approximation of the data, and hence large bandwidth. Choosing λ small allows for large control and forces the error in the approximation of the data to be small, and hence small bandwidth. The bandwidth is particularly apparent in the limit when the spline is replaced by the linear filter; see [101] for numerous examples.

The difficulty with forcing the empirical distribution to reflect the small deviations in the probability distribution is that it increases the number of subintervals that must be used and hence increases the risk of numerical instability in the construction of the spline. We attack that problem directly in the next problem.

Problem 3: Recursively handle many points and modify the estimated distribution.

In [55], there is developed a method of recursively generating splines to avoid the problem of using many points. A similar derivation was done in Chapter 6, we used a sequence of cost functions of the form

$$J_k(u, x_0) = \lambda \int_0^T u_k^2(t)dt + g(k) \int_0^T (y_k(t) - y_{k-1}(t))^2 f(t)dt$$
$$+ \sum_{i=1}^N w_i(y_k(t_i^k) - \alpha_i^k)^2 + (x_0^k)'Q(x_0^k).$$

In this cost function, we see that previous data are stored in terms of the spline. Then the new iteration is penalized both for deviations from the previous estimates and for deviations from the new data.

In the cost function, it is convenient to insist that the weight function $f(t)$ be positive on the interval $[0, T]$, but it can be made very small on subintervals where the distribution needs to be modified. The weight function $g(k)$ is a design parameter to be chosen. It reflects the relative importance of the past splines and the current data. In [55] it was chosen to be k^2, but this was taken to reflect the fact that the data were not going to change dramatically on subsequent revolutions and this choice increases the rate of convergence.

7.5.2 Estimating Exponential Distributions

The problems described above have two inherent difficulties that must be faced; the integral must be equal to 1 and the distribution function must be positive. This difficulty can be overcome by focusing on a class of distributions that are somewhat more specialized but still encompass a large number of examples, namely, the exponential distributions.

In order to estimate such an exponential distribution, $e^{f(x)}$, we take the logarithm and are then faced with the problem of the estimation of the func-

tion $f(x)$. There are no restrictions on the sign of $f(x)$, and thus one difficulty is removed.

However, difficulty does arise. We must still insist on the property that

$$\int_{-\infty}^{\infty} e^{f(x)}\,dx = 1.$$

This is a nonlinear constraint, and the Hilbert space setting is not suitable for such constraints. Some of the relevant work in the area includes
[9],[13],[15]. However, it is at least theoretically possible to move the problem into a Banach space setting and still solve it.

One way to avoid this difficulty is quite artificial but has its merits. The
data used to estimate f come from a finite interval, and yet it is natural
to define exponential distributions on the infinite interval. Let us suppose
that the data for determining f come from the finite interval $[-T, T]$. We
determine f using standard smoothing spline techniques so that we know
$f(-T)$, $f(T)$, and the first derivative of f at the endpoints. We extend the
definition of f to the entire line by defining the extension to be $f(t - T)$ to
the right of $T + \epsilon$ and by $f(T - t) + b$ to the left of $T - \epsilon$. This extension is
discontinuous if $\epsilon = 0$, but for positive ϵ we can use a partition of unity to
connect f and the line segments in a smooth manner.

Now, the f constructed is smooth, and hence so is its exponential. Thus
we can choose the two additional parameters to make the total integral equal
to 1. With this approach we avoid the nonlinear constraint but add a quite
artificial extension. There are two major problems here. The first is that
of incorporating nonlinear constraints. This problem is an active area of
research in control theory and arises when it is necessary to construct splines
on nonlinear surfaces. The second problem is how to extend the construction
of f to the entire line in a manner that is not entirely artificial. Both of these
problems remain to be solved.

SUMMARY

In this chapter, we added a monotonicity constraint to the basic smoothing
spline problem. As a result, we could no longer apply the Hilbert space
techniques, but still, using directional derivatives, found that for nilpotent
systems the optimal solution is piecewise polynomial. For the particular
case of $n = 2$, the problem was completely solved using dynamic programming, resulting in monotone smoothing, cubic splines. The discussion of
such continuous constraints was moreover extended to the important problem of approximating probability distributions from data.

Chapter Eight

SMOOTHING SPLINES AS INTEGRAL FILTERS

In this chapter we construct an integral filter and show that the smoothing spline is the natural approximation of this filter. The construction relies heavily on linear-quadratic optimization theory and the associated theory of Hamiltonians and Riccati transforms. Also, we will show how the bandwidth of the filter can be controlled directly by the smoothing parameter and hence we will conclude that smoothing splines can have a very narrow bandwidth, which is good for picking up local behaviors of the data set.

8.1 SMOOTHING CONCEPTS

A basic problem in statistics is the following. Consider a data set

$$D = \{(t_i, \alpha_i) : i = 1, \ldots, N\},$$

which, for convenience, we will assume is comprised of one-dimensional data, although this is really not technically necessary. We are given some class of functions, which may be presented parametrically or nonparametrically. For example, we could be given all lines of the form $y = ax + b$ (parametrically), or the space of all polynomials or smooth functions (nonparametrically).

No matter if the problem is parametric or nonparametric, the basic idea is to, for each point, define a residue. For example,

$$r_i(a, b) = (\alpha_i - at_i - b)^2$$

provides a simple "distance" from the line to the point. Then a function

$$R(a, b) = \sum_{i=1}^{N} r_i(a, b)$$

can be minimized to select a unique choice of a and b, and hence a fixed line from the set of all parameterized lines. (Note that the line may not be unique.) The resulting line "smooths" the data in the sense that the data are now replaced by a line. Whether this is a "good" smoothing depends on the nature of the set of residues and the definition of "good." We could just as easily have used the set of all fourth-order polynomials or any other parameterized set of functions. These techniques are basic tools in statistics and are used extensively in applications–even when they don't apply.

Another class of smoothing techniques are the so called kernel methods. There the discrete data are replaced by a function, often a step function, and the function is integrated against a kernel, that is,

$$y(t) = \int_{-\infty}^{\infty} f(s)k(s,t)ds.$$

Under very weak conditions on the kernel $k(s,t)$, the curve $y(t)$ is smooth. A typical kernel would be a characteristic function

$$k(s,t) = \left\{ \begin{array}{ll} 0, & s < t, \\ 1, & t < s < t+1, \\ 0, & t+1 < s. \end{array} \right.$$

A more sophisticated kernel is given by

$$k(s,t) = \frac{1}{\sqrt{2\pi}\sigma} e^{\frac{-(t-s-\mu)^2}{2\sigma^2}}.$$

Note that what both kernels are doing is averaging the data in particular ways: The first over a finite interval and the second over the entire line. If $k(s,t)$ is a sum of point masses, then, from this viewpoint, we recover weighted moving averages,

$$d_i = \sum_{k=i-\tau}^{i+\tau} w_k \alpha_k.$$

We will see that smoothing splines can be viewed either from the viewpoint of finding the function from a class that best fits the data or as a kernel smoother. Both concepts are useful. We begin by recalling some basic constructions.

Let the cost functional be given by

$$J(u) = \lambda \int_0^T u^2(t)dt + \sum_{i=1}^{N} w_i(y(t_i) - \alpha_i)^2,$$

and we minimize this functional subject to the, by now quite familiar, constraint

$$\dot{x} = Ax + bu, \quad y = cx, \quad x(0) = x_0.$$

This is essentially the cost function that was used by Grace Wahba, and by replacing the constraint with

$$y(t) = ce^{At}x_0 + \int_0^t ce^{A(t-s)}bu(s)ds$$

we get a well-defined member of the Hilbert space $L_2[0,T]$ (which differentiation would not give). Also, since we can embed the initial condition into the data, we will assume that the initial data $x_0 = 0$.

In statistics, the w_i are usually considered to be design parameters and most often are taken to be identically equal to $1/N$. The parameter λ is considered to be important. Intuitively, λ is designed so that the residues have good statistical properties. Ideally, the residues should be identically independently and normally distributed. This is of course a bit much to hope for when using real data. Wahba describes the process in great detail.

It is a well understood in the folklore of statistics that polynomial smoothing splines act as smoothing filters on noisy data, and that they are in some sense band-limited, that is, a small change in one data point has a primary effect on the spline only in a neighborhood of that point. This was studied explicitly by B. Silverman in [87] for the cubic spline. In this chapter we construct an approximate linear filter for control theoretic splines. The construction is based on linear-quadratic optimal control and related filtering and tracking results. This result shows that smoothing splines can be considered as a nonparametric smoother, and we have seen that they can also be viewed parametrically when we consider the basis functions $\ell_i(s)$ as in previous chapters.

We use a quadratic cost function that contains the function representing the data to be approximated; the cost function is minimized subject to the constraints of the control system that is being used to generate the approximating curve.

The main contributions of this chapter are (1) to show that the control theoretic smoothing splines are a discrete approximation of an integral linear filter; (2) to obtain an explicit, well-motivated linear filter [87]; (3) to show that the smoothing parameter controls bandwidth and hence can be used to gain long-term information, or can be used to control the degree of approximation of the smoothing spline to the discrete numerical data; (4) to show that splines are in fact "local" approximations rather than "global" ones as is emphasized in the numerical literature; and (5) to show that while splines are not causal they only depend on the next few data points, not on the entire future.

8.2 SPLINES FROM STATISTICAL DATA

In this section we state the basic assumptions needed and formulate the problems to be solved. For the sake of completeness, we recall the control used to generate the smoothing spline. The complete derivation is given in Chapter 4.

We consider (again) the control system

$$\dot{x} = Ax + bu, \tag{8.1}$$

$$y = cx, \tag{8.2}$$

where we further assume that the system is controllable and observable. Because we are primarily interested in approximation rather than control, we make the assumption that

$$cb = cAb = \cdots = cA^{n-2}b = 0, \tag{8.3}$$

which gives us the maximal smoothness at the data points.

We assume to be given a data set of the form

$$D_N = \{(t_{i,N}, \alpha_{i,N}) \mid 0 < t_{1,N} \le t_{2,N} \le \cdots \le t_{N,N} < T\}, \tag{8.4}$$

and we make no assumptions about how the data were generated. That is, we do not necessarily assume that the data are generated by a system of the form $dx = Fx\,dt + b\,dW$, where dW is a probabilistic measure. We assume that T is fixed and finite, for without this assumption the integrals do not exist.

We assume that there exists a function $g_N \in C[0,T]$ such that $\alpha_{i,N} = g_N(t_{i,N}) + \epsilon_{i,N}$, where $\epsilon_{i,N}$ is a symmetrically distributed random variable. Let $S_N(t)$ be a smooth, piecewise polynomial function such that

$$S_N(t_{i,N}) = \alpha_{i,N},$$

that is, S_N is an interpolating polynomial spline function. We assume that the data are such that there exists a function $f \in L_2[0,T]$ such that

$$\lim_{N \to \infty} \|f - S_N\|_2 = 0, \tag{8.5}$$

and we further assume that there exists a function $g \in C[0,T]$ such that the sequence $\{g_N\}$ converges uniformly to g. These assumptions are evidently weaker than assuming that the data are generated by the output of a dynamical system driven by noise.

We finally assume as given a cost function of the form

$$J_N(u) = \lambda \int_0^T u^2(t)\,dt + \sum_{i=1}^N w_{i,N}(y(t_{i,N}) - \alpha_{i,N})^2. \tag{8.6}$$

We allow $u \in L_2[0,T]$.

The choice of λ in this formula is very important. In the statistical literature, the choice of λ is governed by the need for the differences between the
spline and the data to be iid normal, or that the approximation to the data
be close. We will see that the choice of λ determines the bandwidth of the
filter, and hence whether the filter picks up the long-term behavior of the
data or if it closely approximates the data points. We assume that $u_{i,N} > 0$
for all indices i, N and a critical assumption is that, for every $h \in C[0,T]$,

$$\lim_{N \to \infty} \left| \int_0^T h(t)dt - \sum_{i=1}^N w_{i,N} h(t_{i,N}) \right| = 0, \tag{8.7}$$

that is, the sampling times and weights form a convergent quadrature algorithm. This assumption is satisfied as long as the weights are chosen to
be the weights associated with a convergent quadrature scheme. (See, e.g.,
[23].)

We define a second cost function in terms of the function f of (8.5):

$$J(u) = \int_0^T \lambda u^2(t) + (y(t) - f(t))^2 dt, \tag{8.8}$$

and we can now formulate two optimal control problems. The first will
produce the control that drives the output of the linear system to the control
theoretic smoothing spline; the second will produce an approximation to the
spline function and will be the object of interest for this chapter.

Problem 8.1

$$\min_{u \in L_2[0,T]} J_N(u)$$

subject to the constraints of the system (8.1) and (8.2).

Problem 8.2

$$\min_{u \in L_2[0,T]} J(u)$$

subject to the constraints of the system (8.1) and (8.2).

As before, we define the function $\ell_t(s)$ as

$$\ell_t(s) = \begin{cases} ce^{A(t-s)}b, & t > s, \\ 0, & t \leq s. \end{cases} \tag{8.9}$$

As already shown, the optimal solution to J_N must be of the form

$$u(s) = \sum_{i=1}^{N} \tau_i \ell_{t_i,N}(s),$$

where τ is the solution of

$$(\lambda W_N^{-1} + G)\tau = \alpha_N, \tag{8.10}$$

where G is the Grammian associated with the ℓ_{t_i}.

Let $u_N(t)$ be the unique solution to Problem 8.1 and let u be the unique solution to Problem 8.2. It was shown in Chapter 5 that the sequence $\{u_N(t)\}$ converges to u in a pointwise manner. Thus the solution of Problem 8.2 is an approximation to the control theoretic spline of Problem 8.1. A major goal in this chapter is to find a solution to Problem 8.2 such that

$$y(t) = \int_0^T k(t,s)f(s)ds. \tag{8.11}$$

To simplify the exposition, we will assign $w_i = 1$ in the rest of the chapter.

8.2.1 Reduction to an Operator Equation

In this subsection, we examine some of the properties of the filter from the somewhat formal view of operator theory. Basic properties of the filter are obtained in this manner along with existence and uniqueness.

For typographical convenience define an operator

$$L_t(u) = \int_0^T \ell_t(s)u(s)ds. \tag{8.12}$$

It is clear that $y(t) = L_t(u) + ce^{At}x_0$ and that by replacing the function f by $f(t) - ce^{At}x_0$ the problem remains unchanged. We will thus assume, without loss of generality, that $x_0 = 0$.

We begin by calculating the Gateaux derivative, $DJ(u;w)$. We have, after a straightforward calculation,

$$DJ(u;w) = 2\int_0^T [\lambda u(t)w(t) + (L_t(u) - f(t))L_t(w)] \, dt, \tag{8.13}$$

and calculating the second derivative with respect to u and evaluating at w, we have

$$D^2(u)(w) = 2\int_0^T \left[\lambda w^2(t) + L_t(w)^2\right] dt. \tag{8.14}$$

From this we see that the second derivative is nonnegative and is 0 if and only if $w(t) = 0$. Thus the functional is convex and hence has a unique minimum.

We now return to (8.13) and set it equal to zero to obtain a necessary and sufficient condition for optimality. After some manipulation, we have

$$0 = \int_0^T [\lambda u(t)w(t) + (L_t(u) - f(t))L_t(w)]\, dt$$

$$= \int_0^T \left[\lambda u(s) + \int_s^T (L_t(u) - f(t))\ell_t(s)dt \right] w(s)ds.$$

Now, this expression is 0 for all w, and hence we have that the optimal u satisfies the integral equation

$$\lambda u(s) + \int_s^T (L_t(u) - f(t))\ell_t(s)dt = 0. \tag{8.15}$$

Multiplying this expression by $\ell_t(s)$ and integrating, we have

$$\lambda y(t) + \int_0^t \ell_t(s) \int_s^T (y(r) - f(r))\ell_r(s)drds = 0,$$

and, after a little reorganization,

$$\lambda y(t) + \int_0^t \int_s^T \ell_t(s)\ell_r(s)y(r)drds = \int_0^t \int_s^T \ell_t(s)\ell_r(s)f(r)drds. \tag{8.16}$$

We now define the operator K as

$$K(g) = \int_0^t \int_s^T \ell_t(s)\ell_r(s)g(r)drds \tag{8.17}$$

for $g \in L_2[0,T]$. Now for every $g \in L_2[0,T]$, $K(g)$ is smooth and hence in $L_2[0,T]$. Rewriting (8.16) we have

$$(\lambda I + K)(y) = K(f). \tag{8.18}$$

Lemma 8.3 *The operator K is self-adjoint.*

Proof. We prove the lemma by direct calculation. After substitution and interchanging the order of integration, we have

$$\langle w, Ku \rangle = \int_0^T w(t)K(u)(t)dt$$

$$= \int_0^T \left[\int_s^T w(t)\ell_t(s)dt \right] \left[\int_s^T \ell_r(s)u(r)dr \right] ds$$

$$= \langle Kw, u \rangle.$$

■

We now decompose K as the sum of two operators by changing the order of integration. An elementary calculation shows that

$$K(u) = \int_0^t \left[\int_0^r \ell_t(s)\ell_r(s)ds \right] u(r)dr + \int_t^T \left[\int_0^t \ell_t(s)\ell_r(s)ds \right] u(r)dr.$$
(8.19)

Now, define operators F and B (forward and backward) as

$$F(u) = \int_0^t \left[\int_0^r \ell_t(s)\ell_r(s)ds \right] u(r)dr$$
(8.20)

and

$$B(u) = \int_t^T \left[\int_0^t \ell_t(s)\ell_r(s)ds \right] u(r)dr.$$
(8.21)

Note that F and B are bounded and hence K is bounded. Also note that from the proof of the lemma the operator K is positive. Thus the spectrum of K is bounded below by 0; hence the spectrum of $I + K$ is bounded away from 0 and the operator $I + K$ is injective.

Lemma 8.4 *For $\lambda > 0$, the operator $\lambda I + K$ is one-to-one and onto.*

Proof. It only remains to prove that $I + K$ is onto. Suppose otherwise. Then there exists a function $x \in L_2[0, T]$ such that, for all $y \in L_2[0, T]$, $\langle x, (I + K)(y) \rangle = 0$. We use the fact that the operator is self-adjoint to conclude that $(I + K)x = 0$. This is equivalent to the fact that x is the unique solution to the optimal control problem with cost function

$$J(u) = \int_0^T \left[u^2(t) + y^2(t) \right] dt,$$

subject to the constraint of the system defined by 8.1 and 8.2. However, it is easy to see that the optimal control is identically zero and hence that the corresponding $y(t)$ is identically zero. Thus we conclude that $x = 0$, and hence $\lambda I + K$ is onto. ■

We can thus solve (8.18) to obtain

$$y = (\lambda I + K)^{-1}K(f).$$
(8.22)

In the next section we will explicitly construct a representation of the operator $(\lambda I + K)^{-1}K$ in terms of an associated Riccati equation.

8.3 THE OPTIMAL CONTROL PROBLEM

In this section we return to the optimal control problem to obtain a different representation of the operator K. With the representation we will obtain, we can examine the detailed properties of the operator.

We return to (8.15). This representation of the optimal control can be rewritten as

$$u(t) = -\int_s^T \frac{1}{\lambda} c e^{A(t-s)} b(y(t) - f(t))dt,$$

and in this form we see that it is in dynamic feedback form and is to be fed back through the system adjoint to the original system (8.1) and (8.2). Our first goal is to explicitly write out the relationship between the system and its adjoint.

We begin by letting

$$g(s) = \int_s^T e^{A^T(t-s)} c^T (y(t) - f(t))dt, \qquad (8.23)$$

where we have replaced $ce^{A(t-s)}$ by its transpose. We calculate the derivative of g to obtain

$$\begin{aligned}
\dot{g}(s) &= -A^T g(s) - c^T (y(s) - f(s)) \\
&= -A^T g(s) - c^T cx + c^T f(s),
\end{aligned}$$

where we have used the fact from (8.2) that $y = cx$.

We now see that

$$u(s) = -\frac{1}{\lambda} b^T g(s). \qquad (8.24)$$

From 8.1 and 8.2 we have

$$\begin{aligned}
\dot{x}(s) &= Ax(s) + bu(s) \\
&= Ax(s) - \frac{1}{\lambda} bb^T g(s).
\end{aligned}$$

From the definition of g, we have

$$g(T) = 0$$

and

$$x(0) = x_0.$$

Writing this in the more conventional form of a forced Hamiltonian system, we have

$$
\frac{d}{dt} \begin{pmatrix} x \\ g \end{pmatrix} = \begin{pmatrix} A & -\frac{1}{\lambda}bb^T \\ -c^T c & -A^T \end{pmatrix} \begin{pmatrix} x \\ g \end{pmatrix} + \begin{pmatrix} 0 \\ c^T \end{pmatrix} f, \qquad (8.25)
$$

with boundary conditions

$$
g(T) = 0 \quad \text{and} \quad x(0) = x_0. \qquad (8.26)
$$

Thus, from the solution of this problem we can explicitly construct the approximate spline $y(t)$. Note that the symplectic matrix is the matrix associated with the Hamiltonian for the optimal control problem

$$
\min_u \int_0^T \left[y(t)^2 + u(t)^2 \right] dt, \qquad (8.27)
$$

subject to the constraint

$$
\dot{x}(t) = Ax(t) + bu(t), \quad y(t) = cx(t).
$$

To solve the two-point boundary value problem we introduce the Riccati transform.

$$
\begin{pmatrix} x \\ w \end{pmatrix} \begin{pmatrix} I & 0 \\ -P(t) & I \end{pmatrix} \begin{pmatrix} x \\ g \end{pmatrix}. \qquad (8.28)
$$

Applying this change of basis to the two-point boundary value problem, we have, after a considerable amount of matrix multiplication

$$
\frac{d}{dt} \begin{pmatrix} x \\ w \end{pmatrix} = \begin{pmatrix} A - \frac{1}{\lambda}bb^T P(t) & -\frac{1}{\lambda}bb^T \\ R(t) & -(A - \frac{1}{\lambda}bb^T P(t))^T \end{pmatrix} \begin{pmatrix} x \\ w \end{pmatrix}
$$
$$
+ \begin{pmatrix} 0 \\ c^T \end{pmatrix} f(t),
$$

where

$$
R(t) = -\dot{P} - PA - c^T c + P\frac{1}{\lambda}bb^T P - A^T P.
$$

We set $R(t) = 0$ and assign it the terminal value $P(T) = 0$. Under the conditions we have imposed of observability and controllability of the original system, this Riccati equation has a unique solution on the interval $[0, T]$. We thus have the following system of equations to solve:

$$
\dot{P} = -PA - c^T c + P\frac{1}{\lambda}bb^T P - A^T P, \quad P(T) = 0, \qquad (8.29)
$$

$$\dot{w} = -\left(A - \frac{1}{\lambda}bb^T P(t)\right)^T w + c^T f, \quad w(T) = 0, \qquad (8.30)$$

$$\dot{x} = \left(A - \frac{1}{\lambda}bb^T P(t)\right) x - \frac{1}{\lambda}bb^T w, \quad x(0) = x_0. \qquad (8.31)$$

We begin by solving and storing the solution of the Riccati equation and substituting this into (8.30). We now have a linear time-varying terminal value problem to solve. Let $\Phi(t, \tau)$ be the solution of

$$\frac{d}{dt}\Phi(t, \tau) = \left(A - \frac{1}{\lambda}bb^T P(t)\right)\Phi(t, \tau),$$

with initial data given by

$$\Phi(\tau, \tau) = I,$$

and let $\Psi(t, \tau)$ be the solution of

$$\frac{d}{dt}\Psi(t, \tau) = -\left(A - \frac{1}{\lambda}bb^T P(t)\right)^T \Psi(t, \tau),$$

with initial data given by

$$\Psi(\tau, \tau) = I.$$

Hence, the solution of (8.30) is given by

$$w(t) = -\int_t^T \Psi(t, \tau)c^T f(\tau)d\tau, \qquad (8.32)$$

and the solution of (8.31) is given by

$$x(t) = \Phi(t, 0)x_0 - \int_0^t \Phi(t, s)\frac{1}{\lambda}bb^T w(s)ds. \qquad (8.33)$$

Concatenating the two solutions gives

$$y(t) = c\Phi(t, 0)x_0 + c\int_0^t \Phi(t, s)\frac{1}{\lambda}bb^T \int_s^T \Psi(s, r)c^T f(r)drds. \qquad (8.34)$$

Changing the order of integration, we have

$$y(t) = c\Phi(t, 0)x_0 + \int_0^t \int_0^r c\Phi(t, s)\frac{1}{\lambda}bb^T \Psi(r, s)c^T ds f(r)dr$$

$$+ \int_t^T \int_0^t c\Phi(t, s)\frac{1}{\lambda}bb^T \Psi(r, s)c^T ds f(r)dr. \qquad (8.35)$$

Thus we have

$$y(t) = c\Phi(t,0)x_0 + \int_0^T k(t,\sigma)f(\sigma)d\sigma, \qquad (8.36)$$

where

$$k(t,\sigma) = \begin{cases} \frac{1}{\lambda}c\int_0^\sigma \Phi(t,\tau)bb^T\Psi(\tau,\sigma)c^T\,d\tau, & 0 \le \sigma \le t, \\ \frac{1}{\lambda}c\int_0^t \Phi(t,\tau)bb^T\Psi(\tau,\sigma)c^T\,d\tau, & t \le \sigma \le T. \end{cases} \qquad (8.37)$$

8.3.1 Simplification of the Formula

We have shown that we can give an explicit expression for the state transition matrix in terms of the system parameters and the solution of the Riccati equation. To solve and store the entire solution of the Riccati equation is very expensive. We will now see that it is sufficient to obtain and store the initial value for the Riccati equation. This simplification is the critical technical step of this section.

We consider the system (8.25), with $x(0) = x_0$ and $g(T) = 0$. By the variation of parameters formula, we obtain

$$\begin{pmatrix} x(t) \\ g(t) \end{pmatrix} = e^{(t-T)H} \begin{pmatrix} x(T) \\ 0 \end{pmatrix} + \int_T^t e^{(t-s)H} \begin{pmatrix} 0 \\ c^T f(s) \end{pmatrix} ds, \quad (8.38)$$

where

$$H = \begin{pmatrix} A & -\frac{1}{\lambda}bb^T \\ -c^Tc & -A^T \end{pmatrix}.$$

This yields the relation between boundary values $x(T)$ and $g(0)$,

$$\begin{pmatrix} x_0 \\ g(0) \end{pmatrix} = e^{-TH} \begin{pmatrix} x(T) \\ 0 \end{pmatrix} + \int_T^0 e^{-sH} \begin{pmatrix} 0 \\ c^T f(s) \end{pmatrix} ds. \quad (8.39)$$

Now, we partition the matrix e^{tH} as follows:

$$e^{tH} = \begin{pmatrix} X_1(t) & X_2(t) \\ Y_1(t) & Y_2(t) \end{pmatrix},$$

where $X_i, Y_i, i = 1, 2$ are $n \times n$ matrices. Due to the semigroup properties, we have the following identities which will be used later:

$$X_1(t-s) = X_1(t)x_1(-s) + X_2(t)Y_1(-s), \qquad (8.40)$$
$$0 = X_1(t)X_2(-t) + X_2(t)Y_2(-t).$$

The unique positive definite solution to the Riccati equation is given by

$$P(t) = Y_1(t-T)X_1(t-T)^{-1}, \qquad (8.41)$$

using the standard Hamiltonian argument.

Then, solving the first block of equations in 8.39 yields

$$x(T) = X_1(-T)^{-1} \left(x_0 + \int_0^T X_2(-s)c^T f(s)ds \right), \qquad (8.42)$$

which, together with (8.38), leads to the following expression

$$x(t) = X_1(t-T)X_1(-T)^{-1} \left(x_0 + \int_0^T X_2(-s)c^T f(s)ds \right)$$

$$+ \int_t^T X_2(t-s)c^T f(s)ds. \qquad (8.43)$$

By the identities from (8.40) and (8.41), we have that $x(t)$ is given by

$$(X_1(t) + X_2(t)P(0))x_0 + \int_0^T (X_1(t) + X_2(t)P(0))X_2(-s)c^T f(s)ds$$

$$- \int_t^T (X_1(s)X_2(-s) + X_2(t)Y_2(-s))c^T f(s)ds \qquad (8.44)$$

$$= (X_1(t) + X_2(t)P(0))x_0 + \int_0^t (X_1(t) + X_2(t)P(0))X_2(-s)c^T f(s)ds$$

$$+ \int_t^T ((X_1(t) + X_2(t)P(0))X_2(-s) - (X_1(s)X_2(-s)$$

$$+ X_2(t)Y_2(-s))) c^T f(s)ds \qquad (8.45)$$

$$= (X_1(t) + X_2(t)P(0))x_0 + \int_0^t (X_1(t) + X_2(t)P(0))X_2(-s)c^T f(s)ds$$

$$+ \int_t^T (X_2(t)(P(0)X_2(-s) - Y_2(-s))c^T f(s)ds. \qquad (8.46)$$

Finally, we obtain the kernel

$$k(t,\sigma) = \begin{cases} c(X_1(t) + X_2(t)P(0))X_2(-\sigma)c^T, & 0 \le \sigma \le t, \\ cX_2(t)(P(0)X_2(-\sigma) - Y_2(-\sigma))c^T, & t \le \sigma \le T. \end{cases}$$

This kernel is the same as the following explicit formula in terms of the system parameters and the Riccati solution:

$$k(t,\sigma) = \begin{cases} c \begin{pmatrix} I & 0 \end{pmatrix} e^{tH} \begin{pmatrix} I & 0 \\ P(0) & 0 \end{pmatrix} e^{-\sigma H} \begin{pmatrix} 0 \\ I \end{pmatrix} c^T, & 0 \le \sigma \le t, \\[3mm] c \begin{pmatrix} I & 0 \end{pmatrix} e^{tH} \begin{pmatrix} 0 & 0 \\ P(0) & -I \end{pmatrix} e^{-\sigma H} \begin{pmatrix} 0 \\ I \end{pmatrix} c^T, & t \le \sigma \le T. \end{cases}$$

$$(8.47)$$

Furthermore, we have, by a simple observation, that

$$\int_0^\sigma \Phi(t,\tau)\frac{1}{\lambda}bb^T\Phi(\sigma,\tau)^T d\tau = (\; I \quad 0\;)\, e^{tH} \begin{pmatrix} I & 0 \\ P(0) & 0 \end{pmatrix} e^{-\sigma H} \begin{pmatrix} 0 \\ I \end{pmatrix}$$

$$\text{for } 0 \le \sigma \le t,$$

$$\int_0^t \Phi(t,\tau)\frac{1}{\lambda}bb^T\Phi(\sigma,\tau)^T d\tau = (\; I \quad 0\;)\, e^{tH} \begin{pmatrix} 0 & 0 \\ P(0) & -I \end{pmatrix} e^{-\sigma H} \begin{pmatrix} 0 \\ I \end{pmatrix}$$

$$\text{for } t \le \sigma \le T,$$

and

$$\Phi(t,0) = (\; I \quad 0\;)\, e^{tH} \begin{pmatrix} I \\ P(0) \end{pmatrix}.$$

Therefore, the transition matrix $\Phi(t,s)$ is

$$\Phi(t,s) = (\; I \quad 0\;)\, e^{tH} \begin{pmatrix} I \\ P(0) \end{pmatrix} \left((\; I \quad 0\;)\, e^{sH} \begin{pmatrix} I \\ P(0) \end{pmatrix} \right)^{-1}.$$

We now see that in order to explicitly construct the optimal filter, we need only find the initial data for the Riccati equation associated with the optimal control problem of (8.27), and the matrix exponential of the symplectic matrix H.

8.4 THE CUBIC SMOOTHING SPLINE

In this section we consider the most important of the splines, the cubic spline, and construct the explicit linear filter. In numerical analysis, the cubic spline is the spline most commonly used. Recall that it is a piecewise cubic polynomial, and is twice continuously differentiable everywhere. In light of the previous section, we begin by deriving the associated Riccati equation.

Let

$$A = \begin{pmatrix} 0 & 1 \\ 0 & 0 \end{pmatrix}, \quad b = \begin{pmatrix} 0 \\ 1 \end{pmatrix}, \quad c = (\; 1 \quad 0\;).$$

Let

$$J(u) = \int_0^T \left[y(t)^2 + \lambda^{-1}u(t)^2 \right] dt,$$

where λ is a positive constant. We now solve the following problem:

$$\min_{u(t)} J(u)$$

subject to the constraints that

$$\dot{x} = Ax + bu, \quad y = cx.$$

The Hamiltonian matrix associated with this problem is

$$H = \begin{pmatrix} 0 & 1 & 0 & 0 \\ 0 & 0 & 0 & -\frac{1}{\lambda} \\ -1 & 0 & 0 & 0 \\ 0 & 0 & -1 & 0 \end{pmatrix}.$$

8.4.1 Explicit Solutions

Our immediate task is to calculate e^{Ht}. To do so we need the explicit powers of H,

$$H^2 = \begin{pmatrix} 0 & 0 & 0 & -\frac{1}{\lambda} \\ 0 & 0 & \frac{1}{\lambda} & 0 \\ 0 & -1 & 0 & 0 \\ 1 & 0 & 0 & 0 \end{pmatrix}, \quad H^3 = \begin{pmatrix} 0 & 0 & \frac{1}{\lambda} & 0 \\ -\frac{1}{\lambda} & 0 & 0 & 0 \\ 0 & 0 & 0 & \frac{1}{\lambda} \\ 0 & 1 & 0 & 0 \end{pmatrix},$$

$$H^4 = \begin{pmatrix} -\frac{1}{\lambda} & 0 & 0 & 0 \\ 0 & -\frac{1}{\lambda} & 0 & 0 \\ 0 & 0 & -\frac{1}{\lambda} & 0 \\ 0 & 0 & 0 & -\frac{1}{\lambda} \end{pmatrix}.$$

From this calculation, we see that the eigenvalues of H are the fourth roots of $-\frac{1}{\lambda}$, namely,

$$\left(\frac{1}{\lambda}\right)^{\frac{1}{4}} \left(\pm 2^{\frac{-1}{2}} \pm i 2^{\frac{-1}{2}}\right).$$

We now give a different form of $\exp Ht$:

$$
\begin{aligned}
e^{Ht} &= \sum_{k=0}^{\infty} \frac{H^k t^k}{k!} \\
&= \sum_{k=0}^{\infty} \frac{H^{4k} t^{4k}}{(4k)!} + \sum_{k=0}^{\infty} \frac{H^{4k+1} t^{4k+1}}{(4k+1)!} + \sum_{k=0}^{\infty} \frac{H^{4k+2} t^{4k+2}}{(4k)!} \\
&\quad + \sum_{k=0}^{\infty} \frac{H^{4k+3} t^{4k+3}}{(4k+3)!} \\
&= I \sum_{k=0}^{\infty} \frac{(-\frac{1}{\lambda})^k t^{4k}}{(4k)!} + H \sum_{k=0}^{\infty} \frac{(-\frac{1}{\lambda})^k t^{4k+1}}{(4k+1)!} + H^2 \sum_{k=0}^{\infty} \frac{(-\frac{1}{\lambda})^k t^{4k+2}}{(4k+2)!}
\end{aligned}
$$

$$+ H^3 \sum_{k=0}^{\infty} \frac{(-\frac{1}{\lambda})^k t^{4k+3}}{(4k+3)!}$$
$$= f_0(t)I + f_1(t)H + f_2(t)H^2 + f_3(t)H^3.$$

We note that

$$He^{Ht} = f_0 H + f_1 H^2 + f_2 H^3 - \frac{1}{\lambda} f_3 I = f_0' I + f_1' H + f_2' H^2 + f_3' H^3,$$

and hence, by linear independence of the powers of H, we have the differential relations

$$f_0' = -\frac{1}{\lambda} f_3, \quad f_0^{(2)} = -\frac{1}{\lambda} f_2, \quad f_0^{(3)} = -\frac{1}{\lambda} f_1,$$

so that it suffices to find a closed form for f_0.

Lemma 8.5 $f_0(t) = \cosh(\frac{(\frac{1}{\lambda})^{1/4}}{\sqrt{2}} t) \cos(\frac{(\frac{1}{\lambda})^{1/4}}{\sqrt{2}} t).$

Proof. Recall that

$$\cosh t = \sum_{k=0}^{\infty} \frac{t^{2k}}{(2k)!}$$

and

$$\cos t = \sum_{k=0}^{\infty} \frac{(-1)^k t^{2k}}{(2k)!},$$

and hence

$$\cosh t + \cos t = 2 \sum_{k=0}^{\infty} \frac{t^{4k}}{(4k)!}.$$

Using the fact that $(\frac{1}{\sqrt{2}} + i \frac{1}{\sqrt{2}})^4 = -1$, we have that

$$\cosh \left(\frac{1}{\lambda}\right)^{1/4} \left(\frac{1}{\sqrt{2}} + i \frac{1}{\sqrt{2}}\right) t + \cos \left(\frac{1}{\lambda}\right)^{1/4} \left(\frac{1}{\sqrt{2}} + i \frac{1}{\sqrt{2}}\right) t$$
$$= 2 \sum_{k=0}^{\infty} \frac{(-1)^k (\frac{1}{\lambda})^k t^{4k}}{(4k)!}.$$

Now, replacing cosh and cos in the above expression by their exponential representations, we have after a little manipulation

$$\cosh \left(\frac{1}{\lambda}\right)^{1/4} \left(\frac{1}{\sqrt{2}} + i \frac{1}{\sqrt{2}}\right) t + \cos \left(\frac{1}{\lambda}\right)^{1/4} \left(\frac{1}{\sqrt{2}} + i \frac{1}{\sqrt{2}}\right) t$$
$$= 2 \cosh \frac{(\frac{1}{\lambda})^{1/4}}{\sqrt{2}} t \cos \frac{(\frac{1}{\lambda})^{1/4}}{\sqrt{2}} t,$$

which completes the proof. ∎

Using the differential relations, we have, after some (tedious) calculations,

$$f_0(t) = \cosh \frac{\left(\frac{1}{\lambda}\right)^{1/4}}{\sqrt{2}} t \cos \frac{\left(\frac{1}{\lambda}\right)^{1/4}}{\sqrt{2}} t,$$

$$f_1(t) = \frac{\lambda^{1/4}}{\sqrt{2}} \left(\cos \frac{\left(\frac{1}{\lambda}\right)^{1/4}}{\sqrt{2}} t \sinh \frac{\left(\frac{1}{\lambda}\right)^{1/4}}{\sqrt{2}} t + \sin \frac{\left(\frac{1}{\lambda}\right)^{1/4}}{\sqrt{2}} t \cosh \frac{\left(\frac{1}{\lambda}\right)^{1/4}}{\sqrt{2}} t \right),$$

$$f_2(t) = \lambda^{1/2} \sin \frac{\left(\frac{1}{\lambda}\right)^{1/4}}{\sqrt{2}} t \sinh \frac{\left(\frac{1}{\lambda}\right)^{1/4}}{\sqrt{2}} t,$$

$$f_3(t) = \frac{-\lambda^{3/4}}{\sqrt{2}} \left(\sinh \frac{\left(\frac{1}{\lambda}\right)^{1/4}}{\sqrt{2}} t \cos \frac{\left(\frac{1}{\lambda}\right)^{1/4}}{\sqrt{2}} t - \cosh \frac{\left(\frac{1}{\lambda}\right)^{1/4}}{\sqrt{2}} t \sin \frac{\left(\frac{1}{\lambda}\right)^{1/4}}{\sqrt{2}} t \right).$$

Thus we have a closed form representation of e^{Ht}.

$$e^{Ht} = \begin{pmatrix} f_0 & f_1 & \frac{1}{\lambda}f_3 & -\frac{1}{\lambda}f_2 \\ -\frac{1}{\lambda}f_3 & f_0 & \frac{1}{\lambda}f_2 & -\frac{1}{\lambda}f_1 \\ -f_1 & -f_2 & f_0 & \frac{1}{\lambda}f_3 \\ f_2 & f_3 & -f_1 & f_0 \end{pmatrix}.$$

Now, let

$$F(t) = \begin{pmatrix} -f_1 & -f_2 \\ f_2 & f_3 \end{pmatrix} \begin{pmatrix} f_0 & f_1 \\ -\frac{1}{\lambda}f_3 & f_0 \end{pmatrix}^{-1}.$$

To show that $F(t)$ is a solution to the Riccati equation, take the derivative of $F(t)$, using the fact that

$$\frac{d}{dt}e^{Ht} = He^{Ht},$$

and then note that $F(0) = 0$. The particular solution we want is then given by

$$P(t) = F(t - T).$$

We now calculate $F(t)$ explicitly. We make the convention that $S = \sinh(\cdot)$, $s = \sin(\cdot), C = \cosh(\cdot), c = \cos(\cdot)$, for typographical convenience.
First

$$F(t) = \frac{1}{f_0^2 + \frac{1}{\lambda}f_1 f_3} \begin{pmatrix} -f_1 f_0 - \frac{1}{\lambda}f_2 f_3 & f_1^2 - f_2 f_0 \\ f_2 f_0 + \frac{1}{\lambda}f_3^2 & -f_1 f_2 + f_3 f_0 \end{pmatrix},$$

which, after some calculations, yields

$$f_1^2 - f_2 f_0 = \frac{\lambda^{1/2}}{\sqrt{2}}(c^2 S^2 + s^2 C^2)$$

$$f_1^2 - f_2 f_0 = f_2 f_0 + \frac{1}{\lambda} f_3^3$$

$$-f_1 f_0 - \frac{1}{\lambda} f_2 f_3 = \frac{\lambda^{1/4}}{\sqrt{2}}(-SC - sc)$$

$$f_3 f_0 - f_1 f_2 = \frac{\lambda^{3/4}}{\sqrt{2}}(-SC + cs)$$

$$f_0^2 + \frac{1}{\lambda} f_1 f_3 = \frac{1}{2}(C^2 + c^2).$$

We have now found an explicit form for $F(t)$:

$$F(t) = \begin{pmatrix} -\sqrt{2}\lambda^{1/4}\frac{SC - cs}{C^2 + c^2} & \lambda^{1/2}\frac{S^2 + s^2}{C^2 + c^2} \\ \lambda^{1/2}\frac{S^2 + s^2}{C^2 + c^2} & \sqrt{2}\lambda^{3/4}\frac{cs - SC}{C^2 + c^2} \end{pmatrix}.$$

Thus the explicit solution of the Riccati equation is

$$P(t) = F(t - T).$$

Remark: It is interesting to note that this derivation is closely related to the representation theory of the cyclic group of order 8. It is possible to construct the Riccati equation for general polynomial splines of degree $2n - 1$, and the construction is closely related to the representation theory of the cyclic group of order $4n$.

We now construct the explicit filter for the cubic spline and exploit the linear quadratic optimization theory to obtain a simplified form of the operator. We first use the fact that $P(0)$ is approximated by the positive definite solution of the algebraic Riccati equation. The steady state Riccati solution for the cubic spline is the positive definite

$$P = \begin{pmatrix} \sqrt{2}\lambda^{1/4} & \sqrt{\lambda} \\ \sqrt{\lambda} & \sqrt{2}\lambda^{3/4} \end{pmatrix}.$$

The transition matrix is

$$\Phi(t, s) = e^{(t-s)(A - \frac{1}{\lambda}bb^T P)},$$

and consequently the kernel $\hat{k}(t, \sigma)$ is given by

$$
\begin{cases}
\dfrac{\exp\left(-\dfrac{\sigma+t}{\sqrt{2}\lambda^{1/4}}\right)}{2\sqrt{2}\lambda^{1/4}}\left[\left(-2+\exp\left(\dfrac{\sqrt{2}\sigma}{\lambda^{1/4}}\right)\right)\cos\dfrac{\sigma-t}{\sqrt{2}\lambda^{1/4}}+\cos\dfrac{\sigma+t}{\sqrt{2}\lambda^{1/4}}\right. \\
\qquad \left. -\exp\left(\dfrac{\sqrt{2}\sigma}{\lambda^{1/4}}\right)\sin\dfrac{\sigma-t}{\sqrt{2}\lambda^{1/4}}-\sin\dfrac{\sigma+t}{\sqrt{2}\lambda^{1/4}}\right], \qquad 0\le\sigma\le t, \\[2ex]
\dfrac{\exp\left(-\dfrac{\sigma+t}{\sqrt{2}\lambda^{1/4}}\right)}{2\sqrt{2}\lambda^{1/4}}\left[\left(-2+\exp\left(\dfrac{\sqrt{2}t}{\lambda^{1/4}}\right)\right)\cos\dfrac{\sigma-t}{\sqrt{2}\lambda^{1/4}}+\cos\dfrac{\sigma+t}{\sqrt{2}\lambda^{1/4}}\right. \\
\qquad \left. +\exp\left(\dfrac{\sqrt{2}t}{\lambda^{1/4}}\right)\sin\dfrac{\sigma-t}{\sqrt{2}\lambda^{1/4}}-\sin\dfrac{\sigma+t}{\sqrt{2}\lambda^{1/4}}\right], \qquad t\le\sigma\le T.
\end{cases}
$$
$$\tag{8.48}$$

It is known that the feedback matrix $A - \frac{1}{\lambda}bb^T P$ is Hurwitz, that is, the eigenvalues of the matrix lie in the left-half complex plane and they are equal to the left-half plane eigenvalues of the matrix H. The following graphs illustrate the differences between our approximations and the kernel defined by Silverman in [87]. Silverman's approximation is given to the kernel

$$
\kappa(t, \sigma) = \frac{1}{2}\exp\left(-\frac{|t-\sigma|}{\sqrt{2}}\right)\sin\left(\frac{|t-\sigma|}{\sqrt{2}}+\frac{\pi}{4}\right). \tag{8.49}
$$

It is worthwhile to compare Silverman's approximation to our approximation in (8.48). Note that two graphs overlap in the plotted range in Figures 8.1 and 8.3.

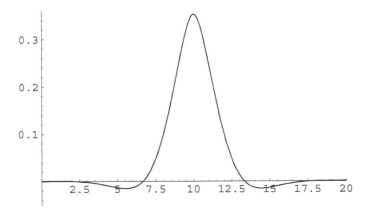

Figure 8.1 The kernels $\kappa(t, \sigma)$ and $k(t, \sigma)$ overlap on the plotted interval.

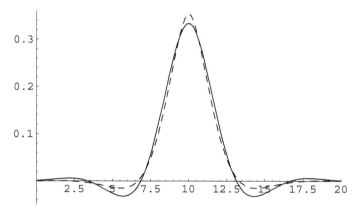

Figure 8.2 The approximated kernel in [87] (solid line) and the kernel $k(t, \sigma)$ (dashed line).

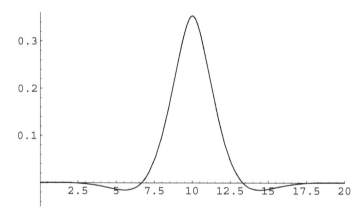

Figure 8.3 The kernels $\kappa(t, \sigma)$ and $\hat{k}(t, \sigma)$ overlap on the plotted interval.

Our approximation is clearly better than Silverman's. However, the improvement is at the expense of complexity of the formula.

8.4.2 Bandwidth of the Kernel

We now turn to a short discussion about the bandwidth of the kernel. The term bandwidth is frequently used in the statistics literature, but is seldom explicitly defined. The definition we use here is that the bandwidth is the interval where one obtains the most information. Concretely, we define the number, β, determined by the solution of the equation

$$\frac{\int_{\max(t-\beta,0)}^{\min(t+\beta,T)} k(t,\sigma)d\sigma}{\int_0^T k(t,\sigma)d\sigma} = 0.9. \tag{8.50}$$

Then 2β is the bandwidth of the kernel $k(t,\sigma)$.

This is a nonlinear equation which can be numerically solved by, for example, the Newton-Raphson method. We illustrate this by some examples involving the approximated kernel. In particular, we compute the bandwidth for (8.48). After integration, we get the denominator (respectively, the numerator) of the left-hand side in (8.50) as follows:

$$\frac{1}{2}e^{-\frac{t+T}{\sqrt{2}\lambda^{1/4}}}\left(-2e^{T}\sqrt{2}\lambda^{1/4}\cos\left(\frac{t}{\sqrt{2}\lambda^{1/4}}\right) - \left(-1+e^{\frac{T}{\sqrt{2}\lambda^{1/4}}}\right)\cos\left(\frac{t-T}{\sqrt{2}\lambda^{1/4}}\right)\right.$$
$$\left. +2\left(e^{\frac{t+T}{\sqrt{2}\lambda^{1/4}}} + \left(-e^{\frac{T}{\sqrt{2}\lambda^{1/4}}}+\cos\left(\frac{T}{\sqrt{2}\lambda^{1/4}}\right)\right)\sin\left(\frac{t}{\sqrt{2}\lambda^{1/4}}\right)\right)\right);$$

$$\frac{1}{2}e^{-\frac{3t}{\sqrt{2}\lambda^{1/4}}}\left(2e^{\frac{3t}{\sqrt{2}\lambda^{1/4}}} - e^{\frac{-\beta+t}{\sqrt{2}\lambda^{1/4}}}\left(-1+e^{\frac{\sqrt{2}\beta}{\lambda^{1/4}}}+2e^{\frac{\sqrt{2}t}{\lambda^{1/4}}}\right)\cos\left(\frac{\beta}{\sqrt{2}\lambda^{1/4}}\right)\right.$$
$$\left. -(e^{\frac{-\beta+t}{\sqrt{2}\lambda^{1/4}}}+e^{\frac{\beta+t}{\sqrt{2}\lambda^{1/4}}})\sin\left(\frac{\beta}{\sqrt{2}\lambda^{1/4}}\right) + e^{\frac{\beta+t}{\sqrt{2}\lambda^{1/4}}}\sin\left(\frac{\beta-2t}{\sqrt{2}\lambda^{1/4}}\right)\right.$$
$$\left. +e^{\frac{-\beta+t}{\sqrt{2}\lambda^{1/4}}}\sin\left(\frac{\beta+2t}{\sqrt{2}\lambda^{1/4}}\right)\right).$$

Solving the equation in some special cases, we obtain $\beta_{T,\lambda}$ for

$$\beta_{10,0.01} = 0.547334, \quad \beta_{10,0.1} = 0.968005, \quad \beta_{10,10} = 3.22425.$$

Remark: We have done an explicit construction for the cubic spline. The same construction can be done for arbitrary polynomial splines and for torsion splines. The limiting difficulty is the explicit calculation of the exponential of the Hamiltonian matrix. However, for most data sets a 4×4 Hamiltonian suffices. It is usually possible to construct the exponential explicitly in these cases. Higher dimensions are problematic.

SUMMARY

In this chapter, we showed that smoothing splines with discrete data lead to a problem with continuous data, and that the solution to the continuous problem leads naturally to an integral filter. This filter is given as a function of the solution to a fixed optimal control problem. The kernel of the filter is obtained exactly. We then showed that the kernel can be significantly simplified and that the kernel depends only on the matrix exponential of a well-behaved Hamiltonian matrix and the initial value of a related Riccati equation. Also, the bandwidth of the filter is related to the smoothing parameter, and hence we concluded that smoothing splines can have a very narrow bandwidth, which is good for picking up local behavior of the data set.

Chapter Nine

OPTIMAL TRANSFER BETWEEN AFFINE VARIETIES

The problem of optimally transferring the state of a linear system between affine varieties arises in a number of applications such as path planning and robot coordination. In this chapter, this problem, along with some of its generalizations, is solved as an example of the control theoretic splines framework in action. In particular, we present an algorithm for obtaining globally optimal solutions through a combination of Hilbert space methods and dynamic programming. As a driving application, the problem of leader-based multi-agent coordination is considered.

9.1 POINT-TO-POINT TRANSFER

Even though the main goal of this chapter is to derive an algorithm for transferring the state of the system between affine varieties, we will start by recalling the solution to the point-to-point transfer problem, as previously defined in Chapter 2.

The point-to-point transfer problem involves driving a linear system of differential equations between given boundary states,

$$\begin{cases} \dot{x} = Ax + Bu \\ x(T_0) = x_0, \ x(T_1) = x_1, \end{cases}$$

where $u \in \mathbb{R}^m$ is the control signal, $x \in \mathbb{R}^n$ the state vector, $A \in \mathbb{R}^{n \times n}$, and $B \in \mathbb{R}^{n \times m}$.

The point-to-point transfer should be done in such a way that a cost functional is minimized with respect to the control signal. The cost functional that we choose to study is

$$J(u) = \int_{T_0}^{T_1} u^T(t)u(t)dt,$$

which can be interpreted as the energy of the control signal.

As seen in Chapter 2, this problem can be formulated as the minimum norm problem in an infinite-dimensional Hilbert space (for example, see [62] and [64])

$$\min_{u} \|u\|_{L_2},$$

under the condition that

$$u \in V_{\rho},$$

where V_{ρ} is the affine variety

$$V_{\rho} = \{v \in L_2 \mid \Lambda v = \rho\}.$$

Here, ρ is given by

$$\rho = x_1 - e^{A(T_1 - T_0)} x_0,$$

and the linear operator $\Lambda : L_2 \to \mathbb{R}^n$ is

$$\Lambda u = \int_{T_0}^{T_1} e^{A(T_1 - s)} Bu(s) ds.$$

We recall that this problem has a solution only if $\rho \in \text{Im}(\Lambda)$. Let us assume this is the case. In fact, by assuming that the system is completely controllable, and using the previously established expressions for Λ^* and the controllability Grammian $\Gamma = \Lambda \Lambda^*$, we get

$$u^* = \Lambda^* \Gamma^{-1} \rho = B^T e^{A^T (T_1 - t)} \Gamma^{-1} \left(x_1 - e^{A(T_1 - T_0)} x_0 \right)$$

and

$$J(u^*) = \left(x_1 - e^{A(T_1 - T_0)} x_0 \right)^T \Gamma^{-1} \left(x_1 - e^{A(T_1 - T_0)} x_0 \right).$$

This classic result will now be generalized to the problem of driving the system between multiple affine varieties.

9.2 TRANSFER BETWEEN AFFINE VARIETIES

The goal is to drive the system in such a way that the solution lies on specific affine varieties at given times, as illustrated in Figure 9.1. The dynamics of the system may differ between the affine varieties, as seen in Figure 9.2. [1]

---

[1]Note that this construction is a generalization of the smoothing spline. Here the smoothness at the nodes is not a primary goal and may in fact be lost. However, smoothness at the nodes could be imposed in much the same way that continuity–the focus in this chapter–is imposed.

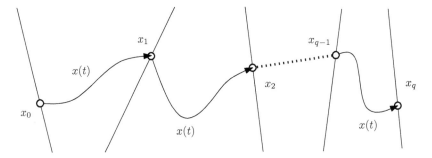

Figure 9.1 Transfer between $q + 1$ affine varieties.

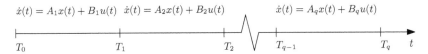

Figure 9.2 System of $q + 1$ affine varieties together with a switched linear system.

Again, as in the point-to-point transfer case, we want to minimize the energy of the control signal. If we denote the affine variety at time T_i by

$$S_i = \{x \in \mathbb{R}^n \mid G_i x = d_i\},$$

we get

$$\min_u J(u)$$

subject to

$$\begin{cases} \dot{x} = A_i x + B_i u \text{ for } t \in (T_{i-1}, T_i), \; i = 1, \ldots, q, \\ x(T_i) \in S_i, \; i = 0, \ldots, q. \end{cases}$$

We will use the solution to the optimal point-to-point transfer problem to formulate a dynamical programming problem from which we can solve for the optimal intersection points on the affine varieties.

Dynamic programming divides the problem into stages with a decision required at each stage. Every stage has a state associated with it. In our case, the stages are represented by the times we are supposed to be on an affine variety, and the state is the state vector at this time. From the point-to-point problem, we know that, given the state at a stage, there is a unique path of minimum cost that takes the system to a specific point at the succeeding stage. So the decision to be made at each stage is where on the affine variety at the succeeding stage we want to end up, given the state we are in at the current stage.

Let $c_i(a, b)$ be the cost of going from state a in stage $i - 1$ to state b in stage i, and let $f_i(a)$ be the minimum cost of going to the affine variety at the final stage, via the intermediate affine varieties, when starting at state a in stage i, as illustrated in Figure 9.3. Let $x_i = x(T_i)$ for $0 \le i \le q$, we then have the following Bellman recursion:

$$f_{i-1}(x_{i-1}) = \min_{x_i \in S_i} (c_i(x_{i-1}, x_i) + f_i(x_i)), \ 0 < i \le q.$$

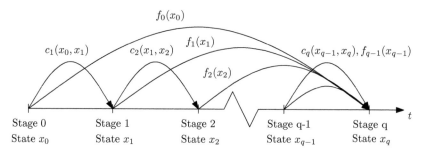

Figure 9.3 Transfer costs in the dynamic programming algorithm.

Our problem can then be reformulated as

$$\min_{x_0 \in S_0} f_0(x_0).$$

Using the cost from the point-to-point problem gives us

$$c_i(a, b) = (b - e^{A_i \Delta T_i} a)^T \Gamma_i^{-1} (b - e^{-A_i \Delta T_i} a),$$

where $\Delta T_i = T_i - T_{i-1}$ and Γ_i is the controllability Grammian associated with the pair (A_i, B_i).

We will assume that the systems are completely controllable. Thus Γ_i is a symmetric positive definite matrix, and so is its inverse Γ_i^{-1}. For notational convenience, let us denote Γ_i^{-1} by Q_i. Then

$$c_i(a, b) = (b - e^{A_i \Delta T_i} a)^T Q_i (b - e^{A_i \Delta T_i} a) = \|b - e^{A_i \Delta T_i} a\|_{Q_i}^2.$$

9.3 TRANSFER THROUGH DYNAMIC PROGRAMMING

Since stage q is the final stage, $f_q(a) = 0$, $\forall a \in \mathbb{R}^n$. With this as the starting condition, we can work our way backward using the Bellman recursion until we get an expression for $f_0(x_0)$. We then minimize $f_0(x_0)$ to get the minimizer x_0^*. By this time we will know the relation between x_0^* and the remaining optimal points on the affine varieties. Once all the optimal points on the affine varieties have been determined, all we need to do is to compute a sequence of point-to-point transfers:

$$
\begin{aligned}
f_{q-1}(x_{q-1}) &= \min_{x_q \in S_q} \{c_{q-1}(x_{q-1}, x_q) + f_q(x_q)\} \\
&= \min_{x_q \in S_q} \|x_q - e^{A_q \Delta T_q} x_{q-1}\|_{Q_q}^2.
\end{aligned}
$$

Now, define the finite-dimensional Hilbert space $\mathcal{H}_q = \mathbb{R}^n$, with inner product

$$
\langle x, y \rangle_{\mathcal{H}_q} = \langle x, y \rangle_{Q_q} = x^T Q_q y
$$

for $x, y \in \mathcal{H}_q$. Since S_q defines an affine variety

$$
V_q^{d_q} = \{x \in \mathcal{H}_q \mid G_q x = d_q\}
$$

in $\mathcal{H}_q$, what we want is to find the $x_q \in V_q^{d_q}$ closest to $p_q = e^{A_q \Delta T_q} x_{q-1} \in \mathcal{H}_q$, that is, to solve

$$
\min_{x_q \in V_q^{d_q}} \|x_q - p_q\|_{\mathcal{H}_q}^2.
$$

Again, according to Hilbert's projection theorem, this problem has a unique optimal solution given by

$$
x_q^* = V_q^{d_q} \cap (V_q^{0\perp} + p_q),
$$

where

$$
V_q^{0\perp} = \{x \mid \langle x, y \rangle_{\mathcal{H}_q} = x^T Q_q y = 0, \ \forall \, y \in V_q^0\}
$$

is the orthogonal complement of the linear subspace

$$
V_q^0 = \{x \in \mathcal{H}_q \mid G_q x = 0\}.
$$

And, since $\mathrm{Im}(G_q^T) = \mathrm{Ker}(G_q)^\perp$, we have

$$
V_q^{0\perp} = \{x \in \mathcal{H}_q \mid \exists \lambda \in \mathbb{R}^{\mathrm{rank}(G_q)} \text{ s.t. } Q_q x = G_q^T \lambda\}.
$$

Now, since the optimal solution is given by $x_q^* = V_q^{d_q} \cap (V_q^{0\perp} + p_q)$, this results in the following linear system of equations for the optimal point:

$$
\begin{cases}
G_q x_q^* = d_q, \\
Q_q(x_q^* - p_q) = Q_q(x_q^* - e^{A_q \Delta T_q} x_{q-1}) = G_q^T \lambda_q,
\end{cases}
$$

or, written in matrix form,

$$P_q \begin{pmatrix} x_q^* \\ \lambda_q \end{pmatrix} = \begin{pmatrix} Q_q p_q \\ d_q \end{pmatrix},$$

where

$$P_q = \begin{pmatrix} Q_q & -G_q^T \\ G_q & 0 \end{pmatrix}.$$

Since the system has a unique solution according to the projection theorem, P_q^{-1} exists. Denote the upper left $n \times n$ matrix of P_q^{-1} by $(P_q^{-1})_{11}$ and the upper right $n \times \text{rank}(G_q)$ matrix of P_q^{-1} by $(P_q^{-1})_{12}$.

This gives

$$x_q^* = H_q x_{q-1} - h_q,$$

where

$$H_q = (P_q^{-1})_{11} Q_q e^{A_q \Delta T_q}$$

and

$$h_q = -(P_q^{-1})_{12} d_q.$$

Proposition 9.6 x_k^* *can be written as an affine function of* x_{k-1}, *that is,* $x_k^* = H_k x_{k-1} - h_k$, $0 < k \le q$.

Proof. We have already seen that this is true for $k = q$. Let us have a look at an arbitrary k, $0 < k < q$. In fact, assume that the claim holds for all i, $k < i \le q$.

$$f_{k-1}(x_{k-1}) = \min_{x_k \in S_k} \{c_k(x_{k-1}, x_k) + f_k(x_k)\},$$

which can be rewritten as

$$\min_{x_k \in S_k} \left\{ \|x_k - p_k\|_{Q_k}^2 + \sum_{j=k+1}^{q} \|F_k^{(j-k)} x_k - p_k^{(j-k)}\|_{Q_j}^2 \right\},$$

where

$$F_k^{(1)} = H_{k+1} - e^{A_{k+1} \Delta T_{k+1}},$$
$$F_k^{(j)} = F_{k+1}^{(j-1)} H_{k+1}, \quad j = 2, \ldots, q - k,$$
$$p_k = e^{A_k \Delta T_k} x_{k-1},$$
$$p_k^{(1)} = h_{k+1},$$
$$p_k^{(j)} = F_{k+1}^{(j-1)} h_{k+1} + p_{k+1}^{(j-1)}, \quad j = 2, \ldots, q - k.$$

Define the finite-dimensional Hilbert space $\mathcal{H}_k = \mathbb{R}^n \times, \cdots, \times \mathbb{R}^n$, with inner product

$$\langle \mathbf{x}, \mathbf{y} \rangle_{\mathcal{H}_k} = \langle x, y \rangle_{Q_k} + \sum_{j=k+1}^{q} \left\langle x^{(j-k)}, y^{(j-k)} \right\rangle_{Q_j},$$

for $\mathbf{x} = (x, x^{(1)}, \ldots, x^{(q-k)})$, $\mathbf{y} = (y, y^{(1)}, \ldots, y^{(q-k)}) \in \mathcal{H}_k$. Define the affine variety $V_k^{d_k}$ as

$$V_k^{d_k} = \{\mathbf{x} \in \mathcal{H}_k \mid G_k x = d_k, \ F_k^{(1)} x^{(1)} = x, \ldots, \ F_k^{(q-k)} x^{(q-k)} = x\}.$$

Then we can write

$$f_{k-1}(x_{k-1}) = \min_{\mathbf{x}_k \in V_k^{d_k}} \{\|\mathbf{x}_k - \mathbf{p}_k\|_{\mathcal{H}_k}^2\}.$$

So the problem is to find $\mathbf{x}_k = (x_k, x_k^{(1)}, \ldots, x_k^{(q-k)}) \in V_k^{d_k}$ closest to $\mathbf{p}_k = (p_k, p_k^{(1)}, \ldots, p_k^{(q-k)}) \in \mathcal{H}_q$. Again, the unique optimal solution is given by $\mathbf{x}_k^* = V_k^{d_k} \cap (V_k^{0\perp} + \mathbf{p}_k)$, where $V_k^{0\perp}$ is the orthogonal complement to the linear subspace V_k^0 given by

$$\left\{ \mathbf{x} \in \mathcal{H}_k \mid G_k x = 0, \ F_k^{(1)} x^{(1)} = x, \ldots, \ F_k^{(q-k)} x^{(q-k)} = x \right\},$$

with $V_k^{0\perp}$ given by

$$\left\{ \mathbf{x} \in \mathcal{H}_k \mid \exists \lambda \in \mathbb{R}^{\mathrm{rank}(G_k)} \right.$$
$$\left. \text{s.t. } Q_k x + \sum_{j=k+1}^{q} F_k^{(j-k)T} Q_j x^{(j-k)} = G_k^T \lambda \right\}.$$

The optimal point $\mathbf{x}_k^* = V_k^{d_k} \cap (V_k^{0\perp} + \mathbf{p}_k)$ is then given by

$$\begin{cases} G_k x_k^* = d_k, \\ F_k^{(1)} x_k^* = x_k^{*(1)}, \\ \quad \vdots \\ F_k^{(q-k)} x_k^* = x_k^{*(q-k)}, \\ Q_k(x_k^* - p_k) \\ \quad + \sum_{j=k+1}^{q} F_k^{(j-k)T} Q_j (x_k^{*(j-k)} - p_k^{(j-k)}) = G_k^T \lambda_k, \end{cases}$$

which is equivalent to

$$\begin{cases} G_k x_k^* = d_k \\ Q_k(x_k^* - p_k) + \\ \quad \sum_{j=k+1}^{q} F_k^{(j-k)T} Q_j (F_k^{(j-k)} x_k^* - p_k^{(j-k)}) = G_k^T \lambda_k. \end{cases}$$

Writing this in matrix form yields

$$P_k \begin{pmatrix} x_k^* \\ \lambda_k \end{pmatrix} =$$

$$\begin{pmatrix} Q_k p_k + \sum_{j=k+1}^{q} F_k^{(j-k)T} Q_j p_k^{(j-k)} \\ d_k \end{pmatrix},$$

with

$$P_k = \begin{pmatrix} Q_k + \sum_{j=k+1}^{q} F_k^{(j-k)T} Q_j F_k^{(j-k)} & -G_k^T \\ G_k & 0 \end{pmatrix}.$$

Let $(P_k^{-1})_{11}$ be the upper left $n \times n$ matrix of P_k^{-1} and $(P_k^{-1})_{12}$ the upper right $n \times rank(G_k)$ matrix of P_k^{-1}. We then have

$$x_k^* = H_k x_{k-1} - h_k,$$

where

$$H_k = (P_k^{-1})_{11} Q_k e^{A_k \Delta T_k},$$

and

$$h_k = -(P_k^{-1})_{12} d_k - (P_k^{-1})_{11} \sum_{j=k+1}^{q} (F_k^{(j-k)T} Q_j p_k^{(j-k)}).$$

So, if the claim holds for all i such that $k < i \le q$, then the claim holds for k. Since the claim holds for $k = q$, by induction, the claim holds for all $k, 0 < k \le q$. ∎

Now we know how to compute $f_0(x_0)$ so all that remains is to find the minimizing x_0:

$$\min_{x_0 \in S_0} \{f_0(x_0)\} = \min_{x_0 \in S_0} \left\{ \sum_{j=1}^{q} \|F_0^{(j)} x_0 - p_0^{(j)}\|_{Q_j}^2 \right\},$$

where

$$F_0^{(1)} = H_1 - e^{A_1 \Delta T_1},$$
$$F_0^{(j)} = F_1^{(j-1)} H_1, j = 2, \ldots, q,$$
$$p_0^{(1)} = h_1,$$
$$p_0^{(j)} = F_1^{(j-1)} h_1 + p_1^{(j-1)}, \ j = 2, \ldots, q.$$

Analogous to the previous construction, we can define the finite-dimensional Hilbert space $\mathcal{H}_0 = \mathbb{R}^n \times, \cdots, \times \mathbb{R}^n$, with inner product

$$\langle \mathbf{x}, \mathbf{y} \rangle_{\mathcal{H}_0} = \sum_{j=1}^{q} \langle x^{(j)}, y^{(j)} \rangle_{Q_j}$$

for $\mathbf{x} = (x^{(1)}, \ldots, x^{(q)})$, $\mathbf{y} = (y^{(1)}, \ldots, y^{(q)}) \in \mathcal{H}_0$. Define the affine variety

$$V_0^{d_0} = \{\mathbf{x} \in \mathcal{H}_0 \mid G_0 x = d_0, \ F_0^{(1)} x^{(1)} = x, \ldots$$
$$F_0^{(q-k)} x^{(q)} = x, \ \text{for some } x \in \mathbb{R}^n\}.$$

We can then write

$$\min_{x_0 \in S_0} \{f_0(x_0)\} = \min_{\mathbf{x}_0 \in V_0^{d_0}} \{\|\mathbf{x}_0 - \mathbf{p}_0\|_{\mathcal{H}_0}^2\},$$

that is, find $\mathbf{x}_0 = (x_0^{(1)}, \ldots, x_0^{(q)}) \in V_0^{d_0}$ closest to $\mathbf{p}_0 = (p_0^{(1)}, \ldots, p_0^{(q)}) \in \mathcal{H}_0$. Again, there exists a unique optimal solution $\mathbf{x}_0^* \in \mathcal{H}_0$, but x_0^* may not be uniquely defined by $\mathbf{x}_0^*$. In fact x_0^* is uniquely defined if and only if $\text{Ker}(G_0) \cap \text{Ker}(F_0^{(1)}) \cap \cdots \cap \text{Ker}(F_0^{(q)}) = \{0\}$.

Using the definition of orthogonality in $\mathcal{H}_0$, and the fact that $\text{Im}(G_0^T) = \text{Ker}(G_0)^\perp$, we get

$$V_0^{0\perp} = \Big\{\mathbf{x} \in \mathcal{H}_0 \mid \exists \lambda \in \mathbb{R}^{\text{rank}(G_0)}$$
$$\text{s.t. } \sum_{j=1}^{q} F_0^{(j)T} Q_j x^{(j)} = G_0^T \lambda \Big\}.$$

The optimal point, $\mathbf{x}_0^* = V_0^{d_0} \cap (V_0^{0\perp} + \mathbf{p}_0)$, is then given by

$$\begin{cases} G_0 x_0^* = d_0, \\ F_0^{(1)} x_0^* = x_0^{*(1)}, \\ \vdots \\ F_0^{(q)} x_0^* = x_k^{*(q)}, \\ \sum_{j=1}^{q} F_0^{(j)T} Q_j (x_0^{*(j)} - p_0^{(j)}) = G_0^T \lambda_0, \end{cases}$$

or equivalently

$$\begin{cases} G_0 x_0^* = d_0, \\ \sum_{j=1}^{q} F_0^{(j)T} Q_j (F_0^{(j)} x_0^* - p_0^{(j)}) = G_0^T \lambda_0. \end{cases}$$

Writing this in matrix form we get

$$
\begin{pmatrix} \sum_{j=1}^{q} F_0^{(j)T} Q_j F_0^{(j)} & -G_0^T \\ G_0 & 0 \end{pmatrix} \begin{pmatrix} x_0^* \\ \lambda_0 \end{pmatrix}
$$
$$
= \begin{pmatrix} \sum_{j=1}^{q} F_0^{(j)T} Q_j p_0^{(j)} \\ d_0 \end{pmatrix}.
$$

As mentioned earlier, the solution to this system may not be unique. However, if we only need to find one solution we can use the Moore-Penrose inverse to get the minimum norm solution.

9.4 A MULTI-AGENT PROBLEM

Let us now apply this result to a particular multi-agent problem, where the network of agents is partitioned into followers and leaders. Such heterogeneous networks were first introduced in [94],[95], through a study of so-called anchor node networks. Following this, a number of issues concerned with *leader-follower networks* have been covered. For instance, controllability was discussed in [79],[51], and the problem of transferring the network between quasi-static equilibrium points was the topic of [52]. The problem of boundary value control was the concern in [40].

The particular example scenario under consideration here is that of *repeated redeployment*, in which the overall mission is specified through a collection of waypoints. These waypoints are, moreover, defined as pairs of interpolation times and subformations, characterizing the desired positions of a subset of the agents at the particular interpolation times. The interpretation here is that additional degrees of control freedom are obtained from the fact that the remaining agents are unconstrained at the interpolation times.

9.4.1 Network Dynamics

Consider N mobile robots, each of which is given by a point in $\mathbb{R}^n$. We will assume that the dynamics associated with each agent are given by $\dot{x}_i = u_i$, $i = 1, \ldots, N$, which means that, along each dimension, the dynamics can be decoupled. Hence, we can, without loss of generality, consider each dimension independently. In other words, let $x_i \in \mathbb{R}$, $i = 1, \ldots, N$, be the position of the ith agent, and $x = (x_1, x_2, \ldots, x_N)^T$ be the aggregated state vector. A widely adopted distributed control strategy for such systems is the so-called *consensus equation*

$$\dot{x}_i = - \sum_{j \in N(i)} (x_i - x_j), \tag{9.1}$$

where $j \in N(i)$ means that there is a connection (i.e., a communication link) between agents i and j.

We will assume that the network topology is static, that is, $N(i)$ does not vary over time. In fact, the consensus equation in (9.1) has been thoroughly studied for static as well as dynamic networks. (A representative sample of some of the highlights in this area of research can be found in [50],[39],[61],[60],[71],[81],[93],[75],[27].)

Algebraic graph theory (see, e.g., [44]) provides us with the tools for analyzing such control strategies: A graph $\mathcal{G} = (V, E)$ consists of a set of nodes $V = \{v_1, v_2, \ldots, v_N\}$, which correspond to the different agents, and a set of edges $E \subset V \times V$, which relates to a set of unordered pairs of agents. A connection exists between agent i and j if and only if $(v_i, v_j) = (v_j, v_i) \in E$; the interpretation here is that $(v_i, v_j) \in E$ if and only if agents i and j have established a communication link between them.

Furthermore, the decentralized control law in (9.1) can be written as

$$\dot{x} = -\mathcal{L}(\mathcal{G})x, \tag{9.2}$$

where $\mathcal{L}(\mathcal{G})$ is the graph Laplacian for the graph $\mathcal{G}$ given by $L(\mathcal{G}) = D(\mathcal{G}) - A(\mathcal{G})$, where $D(\mathcal{G})$ is the degree matrix and $A(\mathcal{G})$ is the adjacency matrix associated with $\mathcal{G}$.

The leader-follower structure of the heterogeneous network is obtained by partitioning the nodes (agents) into leaders and followers, respectively. We will assume that this partitioning is done by assuming that the first $N_f < N$ robots are followers and the remaining $N_l = N - N_f$ robots are leaders, that is, $x = (x_f^T, x_l^T)^T$, where $x_f \in \mathbb{R}^{N_f}$ are the followers' positions and $x_l \in \mathbb{R}^{N_l}$ are the leaders' positions.

The graph Laplacian can then be partitioned as

$$\mathcal{L}(\mathcal{G}) = \left(\begin{array}{cc} \mathcal{L}_f & l_{fl} \\ l_{fl}^T & \mathcal{L}_l \end{array} \right),$$

where $\mathcal{L}_f \in \mathbb{R}^{N_f \times N_f}$, $\mathcal{L}_l \in \mathbb{R}^{N_l \times N_l}$, and $l_{fl} \in \mathbb{R}^{N_f \times N_l}$. Assuming that we can control the velocities of the leader agents directly, we thus get the following dynamics:

$$\dot{x} = \left(\begin{array}{cc} -\mathcal{L}_f & -l_{fl} \\ 0 & 0 \end{array} \right) x + \left(\begin{array}{c} 0 \\ I \end{array} \right) u, \tag{9.3}$$

or $\dot{x} = Ax + Bu$.

9.4.2 Repeated Deployment

What we want to do is drive this system through a collection of waypoints defined as pairs of interpolation times and corresponding desired positions for particular subsets of the agents. This task can be described through a collection of specific affine varieties (defined at the interpolation times) as

$$G_i x(T_i) = d_i, \quad i = 0, \ldots, q, \tag{9.4}$$

where q is the total number of waypoints, the T_i are the interpolation times, and where, for all i, G_i has full rank, $\mathrm{rank}(G_i) \le n$, and $d_i \in \mathbb{R}^{\mathrm{rank}(G_i)}$. (For example, in three dimensions $\{x \mid G_i x = d_i\}$ represents a plane if $\mathrm{rank}(G_i) = 1$, a line if $\mathrm{rank}(G_i) = 2$, or a point if $\mathrm{rank}(G_i) = 3$.)

We want to achieve this repeated transfer between affine varieties while minimizing the control energy expended, that is, while minimizing the quadratic cost functional

$$J(u) = \int_{T_0}^{T_q} u^T(t) u(t) dt. \tag{9.5}$$

9.4.3 Solution for an Example Formation

As an example, consider the situation in which the planar agents interact through a network topology encoded through the graph $\mathcal{G}$, given in Figure 9.4.

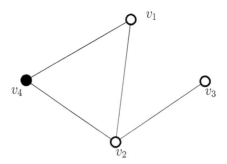

Figure 9.4 A multi-agent network, where agent v_4 is the sole leader.

The graph Laplacian for this system is

$$\mathcal{L}(\mathcal{G}) = \begin{pmatrix} 2 & -1 & 0 & -1 \\ -1 & 3 & -1 & -1 \\ 0 & -1 & 1 & 0 \\ -1 & -1 & 0 & 2 \end{pmatrix}.$$

If we now let $x = (x_1, x_2, x_3, x_4)^T$ be the agent positions in the x-direction and $y = (y_1, y_2, y_3, y_4)^T$ be the agent positions in the y-direction, we obtain the following completely controllable dynamical systems along the two dimensions

$$\dot{x} = -\mathcal{L}x + e_4 u, \quad \dot{y} = -\mathcal{L}y + e_4 v,$$

where e_4 is the unit vector with a 1 in the fourth position and u, v are the scalar control inputs.

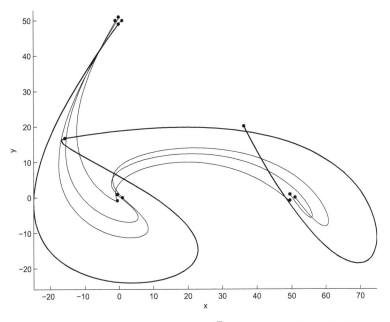

Figure 9.5 Starting at a formation close to $(0\ 50)^T$ at time $t = 0$ the leader (thick curve) maneuvers the followers to the new positions close to $(0\ 0)^T$ at $t = 5$ and close to $(50\ 0)^T$ at $t = 10$. This is done while expending the smallest possible control energy.

Now, the particular repeated redeployment task that we consider is as follows. Given initial positions for all the agents $x(T_0) = x_0$ and $y(T_0) = y_0$, we want to drive the system in such a way that the followers interpolate specific positions at specific times, that is, $x_f(T_i) = x_{fi}$ and $y_f(T_i) = y_{fi}$, $i = 1, \ldots, q$. Since the leader position is unconstrained, we obtain a problem involving affine varieties, and since both the dynamics and the affine varieties are decoupled along the two dimensions, we can solve the problem along each dimension independently. In fact, in Figure 9.5, the optimal solution is given for the minimum energy problem in which all the four agents start "close to" $(x_i, y_i)^T \approx (0, 50)^T$, $i = 1, \ldots, 4$, at time $t = 0$. The leader then moves the followers close to $(x_i, y_i)^T \approx (0,0)^T$, $i = 1, 2, 3$, at $t = 5$ in an optimal fashion, and finally drives them to $(x_i, y_i) \approx (50, 0)^T$, $i = 1, 2, 3$, at time $t = 10$. Figure 9.5 shows this optimal coordinated maneuver.

SUMMARY

In this chapter, we presented the problem of driving a system among different affine varieties. We obtained an algorithm that solves this problem based on dynamic programming and minimum norm optimization in nested Hilbert spaces. A multi-agent coordination problem was solved using this algorithm for computing the optimal leader maneuvers.

Chapter Ten

PATH PLANNING AND TELEMETRY

In this chapter we consider two different applications in which the smoothing spline can be used in a direct manner to solve a seemingly hard problem. The first problem we will consider is the telemetry problem in which global tracking data are measured, and the problem is to reconstruct the noise-free path that generated this data. This will involve producing smoothing splines on spheres, for which we will need to map planar splines onto the sphere using the stereographic projection. The example that we will consider in this context is that of tracking loggerhead sea turtles that can move across vast regions in relatively short amounts of time.

The second example application under consideration in this chapter is how to use the control theoretic splines to generate paths for autonomous robots to track. We will do this in the context of simultaneous collision avoidance and mission progress for multiple aircrafts approaching a landing area. It will turn out that monotone smoothing splines play a key role when solving this particular problem.

10.1 THE TELEMETRY PROBLEM

Radio transmitters are regularly used to track wildlife. A small transmitter is attached to the subject animal and the animal is released. Over the next hours, days, or weeks, depending on the animals and on the particular research problem, signals are recorded from multiple locations at generally nonsynchronized times. These signals are uplinked to a NOAA (National Oceanic and Atmospheric Administration) weather satellite orbiting above Earth, where the signals are preprocessed and stored. Later, as the NOAA satellite passes over a ground station, the information is downlinked.

In order to correctly specify the latitude and longitude of the tracked animal, triangulation techniques are employed, which stresses that a signal for the animal must be available at different satellites at simultaneous times [90], which may or may not always be the case. Furthermore, the achievable accuracy typically depends on the satellite position, with satellites located near the horizon resulting in less accurate position estimates [78]. Thus, it is necessary to construct curves that approximate the data rather than overfitting the noisy data by demanding exact interpolation. Moreover, the signal-to-noise ratio of the individual signals varyies, depending on such factors as transmitter battery charge and temperature. This is the case, for example, when colder temperatures slow down the chemical reactions in the battery, thereby reducing the power available to the transmitter [78]. In order to manage the battery power, signals are normally sent twice a day, on a 13-hour basis and this relatively low sample rate further implies that exact interpolation may not always be desirable.

An additional complication is that in many marine tracking applications, such as migration route mapping, schools of fish or groups of marine animals are tracked. The position of the group is thus given by some combination of clusters, potentially widely spread, of position estimates [72].

These factors, combined with the noise associated with the transmission channel itself, all stress that the algorithms used for producing the migration routes must be approximate in the sense that they should not interpolate exactly through the recorded data points, but rather produce curves on the sphere that do not pay too much attention to outliers [22]. Similar problems arise in many other fields. In animal vision, data are received by the retina, which is roughly spherical. These data points are recorded discretely because of the structure of the retina, and the brain does a very good job of interpolating the data in an approximate fashion [43]. Problems of this type also arise, for instance, when designing trajectories for satellites so that the satellites achieve desired orientations at given times.

The problem of interpolation on differentiable manifolds is a very natural problem and has been studied extensively. In particular, methods have been proposed for deriving interpolating curves on Lie groups (see [11],[20],[48], [74],[99]). These methods are, although elegant in formulation, cumbersome from a computational point of view. One possible remedy to this computational problem is to use Bezier curves, and in particular the use of the De Casteljau algorithm has been proposed [7],[18],[19]. However, the De Casteljau algorithm only works for the exact interpolation case.

In this chapter, we address this problem by projecting the data points onto the plane, using the stereographic projection. It is then straightforward to construct smoothing, generalized splines on the plane, resulting in a trade-off between interpolation and smoothing, as illustrated in [85],[96]. This construction allows the curves to be constructed using optimal control methods for linear systems, established in [30],[91].

10.2 SPLINES ON SPHERES

10.2.1 The Stereographic Projection

The problem under investigation can be stated as follows: Given a set of m data points on the unit sphere, $\xi_i = (z_{1i}, z_{2i}, z_{3i}) \in S^2$, and a set of corresponding times, $t_1 < t_2 < \cdots < t_m$, at which the data were recorded, how do we produce continuously differentiable curves that pass suitably close to these data-points?

If we orient the sphere in such a way that none of the data points coincide with the north pole, that is, if for all $i = 1, \ldots, m$, $(z_{1i}, z_{2i}, z_{3i}) \neq (0, 0, 1)$, then we can always define a stereographic projection that allows us to construct the curves on $\mathbb{R}^2$ instead of trying to construct them on the sphere directly. This strategy makes sense since we are not interested primarily in constructing geodesics, or curves with a similar, special structure, but merely in finding computationally feasible curves that pass close to the data points. These curves should furthermore be at least continuously differentiable everywhere.

The diffeomorphic (rational) stereographic projection, defined on $S^2 - \{(0,0,1)\}$ (the sphere with the "north pole" removed), is given by

$$S(z_1, z_2, z_3) = (y_1, y_2) = \left(\frac{z_1}{1 - z_3}, \frac{z_2}{1 - z_3} \right) \in \mathbb{R}^2, \qquad (10.1)$$

with inverse

$$
\begin{aligned}
S^{-1}(y_1, y_2) &= (z_1, z_2, z_3) \\
&= \left(\frac{2y_1}{y_1^2 + y_2^2 + 1}, \frac{2y_2}{y_1^2 + y_2^2 + 1}, \frac{y_1^2 + y_2^2 - 1}{y_1^2 + y_2^2 + 1} \right) \in S^2.
\end{aligned}
$$
$$(10.2)$$

The time derivative on the sphere is given by

$$
\begin{aligned}
\dot{z}_1 &= (1 - z_3)\dot{y}_1 - z_1(z_1\dot{y}_1 + z_2\dot{y}_2), \\
\dot{z}_2 &= (1 - z_3)\dot{y}_2 - z_2(z_1\dot{y}_1 + z_2\dot{y}_2), \\
\dot{z}_3 &= (1 - z_3)(z_1\dot{y}_1 + z_2\dot{y}_2).
\end{aligned}
$$
$$(10.3)$$

If we rewrite these derivatives in matrix form we get

$$
\frac{d}{dt}
\begin{pmatrix} z_1 \\ z_2 \\ z_3 \end{pmatrix}
=
\begin{pmatrix}
1 - z_3 - z_1^2 & -z_1 z_2 \\
-z_1 z_2 & 1 - z_3 - z_2^2 \\
(1 - z_3)z_1 & (1 - z_3)z_2
\end{pmatrix}
\begin{pmatrix} \dot{y}_1 \\ \dot{y}_2 \end{pmatrix},
\qquad (10.4)
$$

or as $\dot{z} = A(z)\dot{y}$, where $z = (z_1, z_2, z_3)^T$ and $y = (y_1, y_2)^T$.

The second derivative of z is given by

$$\ddot{z} = \left[\frac{d}{dt} A(z) \right] \dot{y} + A(z) \ddot{y}. \tag{10.5}$$

However, from the stereographic projection we can evaluate $\dot{y}$ directly as

$$\dot{y}_1 = \frac{\dot{z}_1}{1 - z_3} + \frac{z_1 \dot{z}_3}{(1 - z_3)^2}$$

and

$$\dot{y}_2 = \frac{\dot{z}_2}{1 - z_3} + \frac{z_2 \dot{z}_3}{(1 - z_3)^2},$$

and in matrix form

$$\begin{pmatrix} \dot{y}_1 \\ \dot{y}_2 \end{pmatrix} = \begin{pmatrix} \frac{1}{1-z_3} & 0 & \frac{z_1}{(1-z_3)^2} \\ 0 & \frac{1}{1-z_3} & \frac{z_2}{(1-z_3)^2} \end{pmatrix} \begin{pmatrix} \dot{z}_1 \\ \dot{z}_2 \\ \dot{z}_3 \end{pmatrix}, \tag{10.6}$$

or symbolically as $\dot{y} = B(z)\dot{z}$. Substituting this into the expression for $\ddot{z}$ in (10.5) gives

$$\ddot{z} = \left[\frac{d}{dt} A(z) \right] B(z)\dot{z} + A(z)\ddot{y}. \tag{10.7}$$

It can be noted that the right-hand side of (10.7) is smooth everywhere except at the singular point $z_3 = 1$.

We now have a characterization of how the dynamics on the plane translates into a dynamics on the sphere, which will prove useful for producing smoothing curves. In the next sections, we will thus cast the wildlife tracking problem as an optimal control problem that can be solved directly in the plane.

10.2.2 Smoothing Splines on the Sphere

Based on the derivations in the previous section, we see that if we control $\ddot{y}_1$ and $\ddot{y}_2$ directly, using a piecewise continuous control signal,

$$\begin{aligned} \ddot{y}_1 &= u_1, \\ \ddot{y}_2 &= u_2, \end{aligned} \tag{10.8}$$

the resulting curves are C^2 everywhere except at points where the input is discontinuous. At these points the curve is C^1.

Substituting (10.8) for $\ddot{y}$ in (10.5) and expressing all variables in terms of $z, \dot{z}$ gives

$$\ddot{z}_1 = -z_1 \frac{\dot{z}_1^2 + \dot{z}_2^2 + \dot{z}_3^2}{1 - z_3} - 2\frac{\dot{z}_1 \dot{z}_3}{1 - z_3} + (1 - z_3 - z_1^2)u_1 - z_1 z_2 u_2,$$

$$\ddot{z}_2 = -z_2 \frac{\dot{z}_1^2 + \dot{z}_2^2 + \dot{z}_3^2}{1 - z_3} - 2\frac{\dot{z}_2 \dot{z}_3}{1 - z_3} + (1 - z_3 - z_2^2)u_2 - z_1 z_2 u_1,$$

$$\ddot{z}_3 = \dot{z}_1^2 + \dot{z}_2^2 - \frac{1 + z_3}{1 - z_3}\dot{z}_3^2 + z_1(1 - z_3)u_1 + z_2(1 - z_3)u_2.$$

If we now introduce a change of variables, $(w_1, w_2, w_3, w_4, w_5, w_6) = (z_1, z_2, z_3, \dot{z}_1, \dot{z}_2, \dot{z}_3)$, we can view the system on the sphere as being directly controlled by the inputs u_1, u_2:

$$\frac{d}{dt} \begin{pmatrix} w_1 \\ w_2 \\ w_3 \\ w_4 \\ w_5 \\ w_6 \end{pmatrix} = \begin{pmatrix} w_4 \\ w_5 \\ w_6 \\ -\frac{w_1}{1-w_3}(w_4^2 + w_5^2 + w_6^2) - \frac{2}{1-w_3}w_4 w_6 \\ -\frac{w_2}{1-w_3}(w_4^2 + w_5^2 + w_6^2) - \frac{2}{1-w_3}w_5 w_6 \\ w_4^2 + w_5^2 - \frac{1+w_3}{1-w_3}w_6^2 \end{pmatrix}$$

$$+ \begin{pmatrix} 0 \\ 0 \\ 0 \\ 1 - w_3 - w_1^2 \\ -w_1 w_2 \\ w_1(1 - w_3) \end{pmatrix} u_1 + \begin{pmatrix} 0 \\ 0 \\ 0 \\ -w_2 w_1 \\ 1 - w_3 - w_2^2 \\ w_2(1 - w_3) \end{pmatrix} u_2.$$

$$(10.9)$$

Note that this system is not controllable on S^2 since the north pole ($z = (0, 0, 1)$) cannot be reached by any control in finite time. On the other hand, the system is controllable when restricted to $S^2 - \{(0, 0, 1)\}$. This is easy to check since, if we want to find a control that maps $a \in S^2$ to $b \in S^2$ in finite time, we use the stereographic projection and map the two points to the plane. We then construct the desired control u^* in the plane and use that control to control the system given by (10.9).

Now, since we do not have any reason to search for geodesics or any other special structure curves, a natural version of the smoothing splines problem is the problem of finding u_1 and u_2 in the Hilbert space of square-integrable functions $L_2[0, T]$ that solve

$$\inf_{u_1, u_2 \in L_2[0,T]} \left\{ \frac{1}{2}\rho \int_0^T (u_1(t)^2 + u_2(t)^2)dt + \frac{1}{2}\sum_{i=1}^m \gamma(z(t_i), \xi_i) \right\}, \quad (10.10)$$

where $0 < t_1 < \cdots < t_m < T$ are the recording times associated with the data points $\xi_1, \ldots, \xi_m$. Furthermore, $\rho > 0$ is a positive smoothing parameter, and $\gamma : S^2 \times S^2 \to \mathbb{R}_+ \cup \{0\}$ is a nonnegative cost associated with letting $z(t_i) \in S^2$ pass close to the data point $\xi_i \in S^2$. In fact, we choose to let $\gamma(z(t_i), \xi_i)$ be defined as

$$\gamma(z(t_i), \xi_i) = \frac{\|\mathcal{S}(\xi_i)\|}{\sum_{j=1}^m \|\mathcal{S}(\xi_j)\|} \|\mathcal{S}(z(t_i)) - \mathcal{S}(\xi_i)\|^2 = \tau_i \|\mathcal{S}(z(t_i)) - \mathcal{S}(\xi_i)\|^2.$$

(10.11)

The reason for this choice of data-fitting cost is threefold. First of all, since small distances between points close to $z = (0, 0, 1)$ are translated into large distances in $\mathbb{R}^2$, when using the stereographic projection, large weights (the τ_i) should be assigned to such points. Second, we do not want the controller to favor points close to the origin. Therefore we need to reduce the weights associated with points close to the south pole of the sphere, $z = (0, 0, -1)$. Third, we want the weights to be normalized, that is, $\sum_{i=1}^m \tau_i = 1$, for convergence reasons to be discussed in the following section.

Finally, the reason for penalizing the control inputs quadratically in (10.10) in this manner is that we get an explicit trade-off between smoothness and curve-fitting. The parameter ρ can be thought of as dictating the degree of data-fitting, as proposed in [96]. However, the real benefit from this way of formulating the wildlife tracking problem is that we can proceed and produce the curves on the plane instead of on the sphere. This significantly simplifies the numerical burden associated with solving the problem in (10.10).

As such, the problem of producing curves on the sphere can be viewed as the problem of generating curves in the plane and then lifting them back to the sphere using the inverse stereographic projection. In this case, we need to use two independent, decoupled linear control systems. Hence the optimization problem becomes

$$\min_{u_1, u_2 \in L_2[0,T]} \left\{ \frac{1}{2}\rho \int_0^T (u_1(t)^2 + u_2(t)^2) dt \right.$$
$$\left. + \frac{1}{2} \sum_{i=1}^m \frac{\|\mathcal{S}(\xi_i)\|}{\sum_{j=1}^m \|\mathcal{S}(\xi_j)\|} \left((y_1(t_i) - \mathcal{S}(\xi_i)_1)^2 + (y_2(t_i) - \mathcal{S}(\xi_i)_2)^2 \right) \right\},$$

(10.12)

subject to

$$\dot{\bar{x}}_i = \begin{pmatrix} 0 & 1 \\ 0 & 0 \end{pmatrix} \bar{x}_i + \begin{pmatrix} 0 \\ 1 \end{pmatrix} u_i, \quad y_i = (1 \quad 0) \bar{x}_i, \quad i = 1, 2, \quad (10.13)$$

where we let $\bar{x}_i$ denote the point $(y_i, \dot{y}_i)$, $i = 1, 2$ in $\mathbb{R}^2$. Furthermore,

$\mathcal{S}(\xi_i)_j$, $j = 1, 2$, denotes the jth component of the stereographic projection of the ith data point.

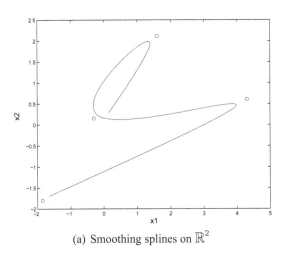

(a) Smoothing splines on $\mathbb{R}^2$

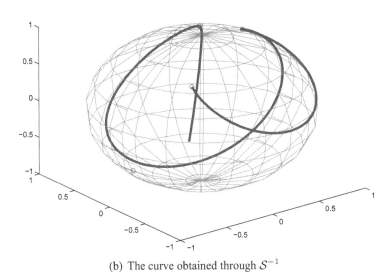

(b) The curve obtained through $\mathcal{S}^{-1}$

Figure 10.1 Smoothing splines on $\mathbb{R}^2$ and S^2. The curve is obtained by projecting the data points (the circles) onto the plane and then lifting the smoothing spline back to the sphere, which results in a curve that passes close to the data points.

If we compare this formulation to the problem in (10.10), we have that

$$\kappa_i = \frac{\|\mathcal{S}(\xi_i)\|}{\sum\limits_{j=1}^{m}\|\mathcal{S}(\xi_j)\|},$$

$$\alpha_{i,j} = \mathcal{S}(\xi_i)_j, \ j = 1, 2,$$

and

$$z(t_i) = \mathcal{S}^{-1}(y_1(t_i), y_2(t_i)).$$

It is clear that this problem is equivalent to that in (10.10), and the reason why inf has been replaced by min is due to the convexity of the problem [64].

A result from using this strategy is shown in Figure 10.1, where planar smoothing splines are lifted back onto the sphere.

10.2.3 Example: Tracking of Marine Animals

As an example of using the proposed method in a wildlife telemetry application, we consider the problem of mapping the route of loggerhead sea turtles along the Atlantic coast of North America. Tracking these turtles in the open sea is done by mounting small transmitters to the backs of the animals. After 8–10 months, the transmitters quit working, and they are typically designed to fall off the turtles at this point [14].

The data received from the turtles contain information about current positions, number of dives taken since last transmission, duration of the most recent dive, and water temperature. The position data (time-stamped longitudinal and latitudinal data) are available through WhaleNet at [63], and given a data point (lat, long), the position on S^2 can be obtained as

$$\begin{pmatrix} z_1 \\ z_2 \\ z_3 \end{pmatrix} = \begin{pmatrix} \cos(\text{lat})\cos(\text{long}) \\ \cos(\text{lat})\sin(\text{long}) \\ \sin(\text{lat}) \end{pmatrix}.$$

In Figure 10.2, the route of the loggerhead sea turtle "Mary-Lee" is tracked and displayed over a three-month period, with a total of 76 data points.

10.3 SPLINES AND BEZIER CURVES

As Bezier curves, and in particular the De Casteljau algorithm, have been proposed as ways to generate (interpolating) splines on spheres, we here say a few words about the connection between splines and Bezier curves.

In fact, we show that Bezier curves can be thought of as the solutions to linear optimal control problems, using results from Hermite interpolation, in combination with traditional linear optimal control. This provides us with a computational view of Bezier curves that differs from the standard De Casteljau algorithm, and it furthermore points out the close relationship between Bezier curves and interpolating dynamic splines.

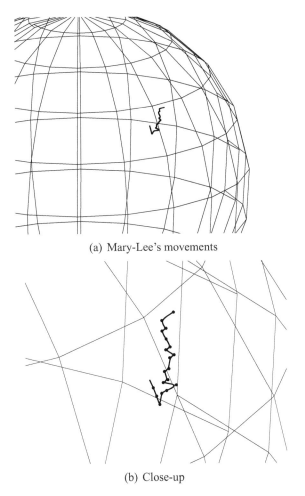

(a) Mary-Lee's movements

(b) Close-up

Figure 10.2 Movements of the loggerhead sea turtle "Mary-Lee" during a three-month period are reconstructed using smoothing splines. Depicted is only a small subset
of the data points.

10.3.1 Bezier Curves

Bezier curves constitute a class of approximating curves in that they are defined using control points, but do not necessarily pass through these control points. Instead, the control points define the shape of the curve as

$$B(t) = \sum_{i=0}^{N} B_{N,i}(t) p_i, \qquad (10.14)$$

where $p_i \in \mathbb{R}^p$, $i = 0, \ldots, N$ are the control points, and

$$B_{N,i}(t) = \binom{N}{i} (1-t)^{N-i} t^i \qquad (10.15)$$

is a Bernstein polynomial. It is immediately clear from (10.15) that the Bezier curves are parameterized by $t \in [0, 1]$.

From (10.15) it furthermore follows that we need $N + 1$ control points in order to define a Bezier curve of degree N. Given $N + 1$ such control points in $\mathbb{R}^p$, the Bezier curves can be established by an iterative algorithm, the De Casteljau algorithm, which produces a single point on the curve for each iteration of the algorithm. The construction is shown in Figure 10.3, where the points $p_0, \ldots, p_4$ are the control points in (10.14). The curve is produced by letting λ sweep $[0, 1]$ as follows. The control points are connected with lines, and new points are defined on those lines at a fraction λ of the distance between the endpoints of the individual lines. In Figure 10.3, those points are $p_{01}, p_{12}, p_{23}, p_{34}$. This procedure is repeated, generating the points $p_{012}, p_{123}, p_{234}$ in the second step and p_{0123}, p_{1234} in the third step. The final point p_{01234} is a point on the Bezier curve, and in this particular case we have that $p_{01234} = B(\lambda)$.

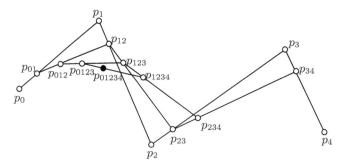

Figure 10.3 The standard construction of a single point on the Bezier curve.

The existence of this computationally inexpensive algorithm is what makes the Bezier curves useful in a number of applications (see, e.g., [19]), and they are used extensively in computer graphics, as well as in such areas as computer aided design. There is a wealth of literature associated with this topic, and the books by Farin [37],[38] are standard references for this subject. However, what will be shown here is that while the DeCasteljau algorithm is elegant, the Bezier curve is in fact a fundamental object from linear control theory.

The question has been raised of whether or not Bezier curves are intrinsically better than the dynamic splines discussed in [68],[85],[91],[96],[100], [103]. We will show that Bezier curves are in fact intimately associated with a linear optimal control problem, and thus that they can be realized as the solution to just such a problem. This means that further research is required before it can be claimed that Bezier curves offer better performance than other spline methods. In producing this result, we will rely heavily on the fact that Bezier curves can be related to certain Hermite interpolation problems, and that in turn these Hermite interpolation problems are in fact linear optimal control problems.

From (10.15), it is easy to calculate the derivatives of a Bezier curve using the differential recursion for the Bernstein polynomials, given by

$$\frac{d}{dt} B_{N,i}(t) = N B_{N-1,i-1}(t) - N B_{N-1,i}(t), \ i = 1, \ldots, N-1, \quad (10.16)$$

and

$$\frac{d}{dt} B_{N,0}(t) = -N B_{N-1,0}(t),$$
$$\frac{d}{dt} B_{N,N}(t) = N B_{N-1,N-1}(t). \quad (10.17)$$

We can thus calculate the derivative of the Bezier curve in (10.14) as

$$\frac{d}{dt} B(t) = N \sum_{i=0}^{N-1} B_{N-1,i}(t)(p_{i+1} - p_i). \quad (10.18)$$

From (10.18) it follows that the derivative is itself a scalar multiple of a Bezier curve, calculated from the differences of the original control points.

It is now straightforward to calculate all the derivatives of the curve. As is shown in [37],[38], these derivatives have a very nice closed form expression in terms of the forward differencing operators Δ_F^k, defined recursively as

$$\Delta_F^k p_j = \Delta_F^{k-1} p_{j+1} - \Delta_F^{k-1} p_j, \ k = 1, 2, \ldots,$$
$$\Delta_F^0 p_j = p_j. \quad (10.19)$$

Thus the kth derivative is given by

$$\frac{d^k}{dt^k} B(t) = \frac{N!}{(N-k)!} \sum_{i=0}^{N-k} B_{N-k,i}(t) \Delta_F^k p_i. \quad (10.20)$$

We now note that only one of the Bernstein polynomials in (10.14) is nonzero at $t = 0$, namely, $B_{N,0}(t)$, with $B_{N,0}(0) = 1$. Hence

$$B(0) = p_0,$$

$$\frac{d}{dt}B(0) = N\Delta_F p_0,$$

$$\vdots \tag{10.21}$$

$$\frac{d^k}{dt^k}B(0) = \frac{N!}{(N-k)!}\Delta_F^k p_0.$$

With this formulation it is possible to calculate the derivatives of the Bezier curve at the two endpoints p_0 and p_N. These two points have a special significance in the Bezier curve construction since they are the only two points that the Bezier curve is guaranteed to interpolate. From (10.21), we see that the interior points determine the derivatives of the Bezier curve at the two endpoints. This suggests that Bezier curves are associated with certain Hermitean interpolation problems. In fact, taking N derivatives gives the two vectors

$$\begin{pmatrix} p_0 \\ N\Delta_F p_0 \\ \vdots \\ N!\Delta_F^N p_0 \end{pmatrix} \in \mathbb{R}^{p(N+1)}, \quad \begin{pmatrix} p_N \\ N\Delta_B p_N \\ \vdots \\ N!\Delta_B^N p_N \end{pmatrix} \in \mathbb{R}^{p(N+1)} \tag{10.22}$$

of derivatives at the endpoints that we would like to interpolate. Here Δ_B is the backward differencing operator, and it should be pointed out that at this point we do not know whether we need to take all N derivatives at the endpoints in order to generate the correct Hermite interpolation problem.

10.3.2 Connections to Linear Systems

We consider a linear system of the form

$$\dot{x} = \mathcal{A}x + \mathcal{B}u, \quad y = \mathcal{C}x, \tag{10.23}$$

where $u, y \in \mathbb{R}^p$ and $x \in \mathbb{R}^{pq}$ for some q to be determined later. Furthermore, we let $\mathcal{A}$ have the form

$$\mathcal{A} = \begin{pmatrix} A & 0 & \cdots & 0 \\ 0 & A & \cdots & 0 \\ \vdots & & \ddots & \vdots \\ 0 & & & \\ & & \cdots & A \end{pmatrix}, \tag{10.24}$$

where A is the $q \times q$ nilpotent matrix

$$
A = \begin{pmatrix}
0 & 1 & 0 & \cdots & 0 \\
0 & 0 & 1 & \cdots & 0 \\
\vdots & \vdots & & \ddots & \vdots \\
0 & \cdots & 0 & 0 & 1 \\
0 & \cdots & & 0 & 0
\end{pmatrix}.
\tag{10.25}
$$

We furthermore let $\mathcal{B}$ be the matrix

$$
\mathcal{B} = (e_q, e_{2q}, \ldots, e_{pq}),
\tag{10.26}
$$

where e_k is the kth unit vector in $\mathbb{R}^{pq}$, with 1 in the kth position. The matrix $\mathcal{C}$ is similarly given as

$$
\mathcal{C} = \begin{pmatrix}
e_1^T \\
e_{1+q}^T \\
\vdots \\
e_{1+(p-1)q}^T
\end{pmatrix}.
\tag{10.27}
$$

The system in (10.23) thus consists of p single-input single-output linear systems in parallel, and hence it suffices to consider the subsystems individually. The algorithm for constructing the Bezier curve is also a coordinate-wise algorithm. Thus we will, without loss of generality, consider the old workhorse, the single-input single-output linear system

$$
\dot{x} = Ax + bu, y = cx,
\tag{10.28}
$$

where A is described above, $b = e_q$ (qth unit vector in $\mathbb{R}^q$), and $c = e_1^T$ (first unit vector in $\mathbb{R}^q$).

A classical result for this type of system is given by the following.

Theorem 10.1 *Let $x(0), x(1) \in \mathbb{R}^q$. Then the control, $u(t)$, that minimizes*

$$
J(u) = \int_0^1 u^2(t)dt
\tag{10.29}
$$

and drives the controllable system

$$
\dot{x} = Ax + bu
\tag{10.30}
$$

from $x(0)$ to $x(1)$, is given by

$$
b^T e^{A^T(1-t)} \left(\int_0^1 e^{A(1-s)} bb^T e^{A^T(1-s)} ds \right)^{-1} (x(1) - e^{A1}x(0)).
\tag{10.31}
$$

We now recall that if A is given in (10.25) we have that the exponential is a polynomial matrix. In fact,

$$
e^{At} = \begin{pmatrix}
1 & t & \frac{t^2}{2!} & \cdots & \frac{t^{p-1}}{(p-1)!} \\
0 & 1 & t & \cdots & \frac{t^{p-2}}{(p-2)!} \\
\ddots & \cdots & \ddots & \ddots & \\
0 & \cdots & & & 1
\end{pmatrix}.
\tag{10.32}
$$

Thus the control in Theorem 10.1 is polynomial, and the degree of the control is $q - 1$.

10.3.3 Two-Point Hermite Interpolation

The classical Hermite interpolation problem is discussed in great detail in almost every elementary numerical analysis book. (In particular, see [22].) We are interested in a specific form of the general problem, namely, the two-point problem. At time 0 we specify k values $a_0, a_1, \ldots, a_{k-1}$, and at time 1 we likewise specify k values $b_0, b_1, \ldots, b_{k-1}$.

The problem we are interested in is to find a polynomial of minimum degree such that $p^{(i)}(0) = a_i$ and $p^{(i)}(1) = b_i$, $i = 0, \ldots, k - 1$. It is easy to see that there exists a unique polynomial of degree less than or equal to $2k - 1$ which satisfies the requirement.

In the previous section we saw that there exists a control law that drives a system from a point in $\mathbb{R}^q$ to another point in $\mathbb{R}^q$ in such a way that the resulting trajectory is polynomial. Using the notation of Theorem 10.1, we see that since $u(t)$ is a polynomial of degree at most $q - 1$, $y(t)$ is in fact also a polynomial of degree at most $2q - 1$. The output $y(t)$ thus satisfies the constraints of the Hermite interpolation problem discussed above, in the case when $k = q$. We can thus conclude that the linear optimal control problem and the special, two-point Hermite problem have the same solution.

We saw previously that, given $N + 1$ points in $\mathbb{R}^p$, the Bezier curve is a polynomial curve of degree N. Since the Bezier algorithm operates at the level of coordinates, we can restrict ourselves to the case $p = 1$, which corresponds to using only the subsystem in (10.28) instead of the full system in (10.23). For the continuation of this section, we consider two cases based on the parity of N.

Case 1: N=2M-1

If we want to produce a solution to the Hermite interpolation problem that has the same degree as the Bezier curve, we need to interpolate between points in $\mathbb{R}^k$ such that $2k - 1 = N$, that is, $k = M$. In order to produce such interpolation points in $\mathbb{R}^M$, we need to compute $M - 1$ derivatives of the Bezier curve.

Using the expression for the derivative of the Bezier curve in (10.20), we have that the derivatives of the curve at $t = 0$ are given by the vector

$$
x(0) = \begin{pmatrix} p_0 \\ (2M - 1)\Delta_F p_0 \\ \vdots \\ \frac{(2M-1)!}{M!}\Delta_F^{M-1} p_0 \end{pmatrix} \in \mathbb{R}^M,
\tag{10.33}
$$

where we have now assumed that $p_0 \in \mathbb{R}$. The corresponding first derivatives at time 1 are given by

$$
x(1) = \begin{pmatrix} p_{2M-1} \\ (2m - 1)\Delta_B p_{2M-1} \\ \vdots \\ \frac{(2M-1)!}{M!}\Delta_B^{M-1} p_{2M-1} \end{pmatrix} \in \mathbb{R}^M.
\tag{10.34}
$$

Now, the linear system of Theorem 10.1 that drives the system (10.28) between $x(0)$ and $x(1)$ produces an output curve of degree $2M - 1$, which is equal to N, the degree of the Bezier curve. However, by the Hermite problem, this curve is unique, and hence the Bezier curve and the curve produced by the linear optimal control law are one and the same.

The case when $N = 2M$ is a bit more involved.

Case 2: N=2M

As before, the degree of the Bezier curve in (10.14) is N, which is equal to $2M$. Let us now proceed by taking M instead of $M - 1$ derivatives in order to get the two endpoints in the Hermite interpolation problem. We get

$$
x(0) = \begin{pmatrix} p_0 \\ 2M\Delta_F p_0 \\ \vdots \\ \frac{(2m)!}{M!}\Delta_F^M p_0 \end{pmatrix} \in \mathbb{R}^{M+1},
\tag{10.35}
$$

$$
x(1) = \begin{pmatrix} p_{2M} \\ 2M\Delta_B p_{2M} \\ \vdots \\ \frac{(2M)!}{M!}\Delta_B^M p_{2M} \end{pmatrix} \in \mathbb{R}^{M+1}.
\tag{10.36}
$$

It is a well-posed problem to construct the polynomial of minimal degree that solves the Hermite interpolation problem. Since we are interpolating in $\mathbb{R}^{M+1}$, we get an upper bound of $2(M+1) - 1 = 2M + 1$ on the degree of the polynomial obtained from the Hermite interpolation. But the Bezier curve also interpolates the same data using a polynomial of degree $2M$. Hence, in this case, the degree of the unique Hermite polynomial is actually $2M$ instead of the generic degree $2M + 1$.

We can also construct a polynomial that interpolates these data using linear control theory. Again, using Theorem 10.1, with the degree of the system being $M + 1$, we construct a polynomial of degree $2M + 1$ that interpolates the data. Again appealing to the uniqueness of the Hermite interpolation problem we conclude that the degree is actually $2M$.

10.3.4 Main Theorem

Based on the observations in the previous two subsections, we have established the following fact that we state as a theorem.

Theorem 10.2 *Let $\{p_i \mid i = 0, \ldots, N\}$ be a set of $N + 1$ points in $\mathbb{R}$, with the corresponding Bezier curve*

$$B(t) = \sum_{i=0}^{N} B_{N,i} p_i, \tag{10.37}$$

where

$$B_{N,i} = \binom{N}{i} (1-t)^{N-i} t^i. \tag{10.38}$$

Let the function $y(t)$ be given by

$$\begin{aligned} \dot{x} &= Ax + bu, \\ y &= cx, \end{aligned} \tag{10.39}$$

where A, b, c are given in (10.28), and where $u(t)$ solves

$$\min_{u} \int_0^1 u^2(t)dt, \tag{10.40}$$

while interpolating

$$x(0) = \begin{pmatrix} p_0 \\ N\Delta_F p_0 \\ \vdots \\ \frac{N!}{(N-m)!}\Delta_F^m p_0 \end{pmatrix} \in \mathbb{R}^{m+1}, \tag{10.41}$$

$$x(1) = \begin{pmatrix} p_N \\ N\Delta_B p_N \\ \vdots \\ \frac{N!}{(N-m)!}\Delta_B^m p_N \end{pmatrix} \in \mathbb{R}^{m+1}. \qquad (10.42)$$

Then $y(t)$ is identical to the Bezier curve $B(t)$, with the choice of $m = \lfloor \frac{N}{2} \rfloor$, with $\lfloor \cdot \rfloor$ being the floor operator.

From Theorem 10.2 it follows that it is possible to use the Bezier curves for constructing interpolating splines. However, the control points must be chosen to ensure continuity of the derivatives at each of the nodes; the procedure for doing this is that in [100]. Hence, the usual claim that Bezier splines are cheaper to compute than interpolating splines to compute is probably not true if it is desired to have a closed form for the resulting spline and there are more than two interpolating nodes. Even though we make no claims that dynamic splines are better than Bezier curves, we have in fact shown that Bezier curves are also a fundamental construction of linear control theory.

10.4 CONFLICT RESOLUTION FOR AUTONOMOUS VEHICLES

We have now seen how to use splines for reconstructing actual paths from sampled data points. How about the opposite problem, to use splines to generate paths? In this section, we will consider this path planning problem in the particular context of conflict avoidance for multiple (possibly autonomous) vehicles, such as Unmanned Aerial Vehicles (UAVs).

To make matters more concrete, the problem we consider here is how to generate paths that lead multiple planar autonomous vehicles to a desired goal state, such as an airport terminal or a robot docking station, in a safe and orderly manner. We do not want to solve the problem of controlling the nonlinear robot dynamics at the same time as we plan collision-free routes, which typically implies that a hierarchical approach is called for.

We assume that each of the m individual robots' states evolves on the smooth manifold $\mathcal{M}$, and that the dynamics is defined by the control system

$$\dot{x}_i(t) = f(x_i(t), u_i(t)), \; x_i(0) = x_{i0}, \; i = 1, \ldots, m,$$

where $x_i(t) \in \mathcal{M}$ and $u_i(t)$ belongs to the set of admissible inputs U. The problem that we investigate concerns driving these robots toward a terminal or docking station in the plane, and for this we associate an output equation

$$y_i(t) = h(x_i(t)) \in \mathbb{R}^2$$

to each robot that projects the robot states onto $\mathbb{R}^2$.

The specifications when landing airplanes are typically given by discretizing the plane, resulting in a planar, Cartesian grid, with respect to which the waypoints are defined. We can, of course, always scale this grid to have it coincide with $\mathbb{Z}^2$, which will be done throughout this section. This grid structure will furthermore be referred to as a cell-partitioning of $\mathbb{R}^2$, where each cell is defined as $\{y \in \mathbb{R}^2 \mid |y(1) - z(1)| \leq 1/2$ and $|y(2) - z(2)| \leq 1/2\}$ for a given $z \in \mathbb{Z}^2$.

In order to formulate the specifications in terms of the robot dynamics, we introduce the mapping $\mathcal{Z} : \mathbb{R}^2 \rightarrow \mathbb{Z}^2$, that maps points in $\mathbb{R}^2$ to the midpoint of the particular cell that the point resides within, that is,

$$\mathcal{Z}(y) = \operatorname{argmin} \|y - z\|_2,$$

taken over $z \in \mathbb{Z}^2$, where $\|\cdot\|_2$ denotes the standard Euclidean norm in $\mathbb{R}^2$. In order to avoid ambiguities we can furthermore always define $\mathcal{Z}$ uniquely on the boundary of each cell.

On $\mathbb{Z}^2$, we can now define the Manhattan metric $\|z\|_M$ as the minimum number of cells that must be traversed in order to reach the cell containing the origin. If we let

$$p_i = \|\mathcal{Z}(h(x_i(0)))\|_M \in \mathbb{Z}^2,$$

we are ready to formulate the multi-agent specifications under consideration here,

1. *Terminal Approach:* $\|\mathcal{Z}(h(x_i(k)))\|_M + 1 = \|\mathcal{Z}(h(x_i(k-1)))\|_M$, $k = 1, \ldots, p_i$, $i = 1, \ldots, m$; and

2. *Collision Avoidance: Given i, $\|h(x_i(t)) - h(x_j(t))\|_2 > \rho$, $\forall j \neq i$, as long as t satisfies $\|\mathcal{Z}(h(x_i(t)))\|_M \neq 0$, where ρ is the desired safety margin.*

It should be noted that by specifying the requirements in this way we have implicitly required an additional property for our solution to exhibit:

3. *Docking:* $\|\mathcal{Z}(h(x_i(p_i)))\|_M = 0$, $i = 1, \ldots, m$.

Instead of explicitly trying to find a controller that satisfies these three specifications directly, we take advantage of the fact that there are already a number of efficient tracking algorithms with provable performance, and we refer to the literature [1],[12],[26],[25],[29],[73],[84] for a treatment of this topic. Thus, we can focus our attention on the problem of generating paths, at various levels in the hierarchy, instead of on tracking these paths.

10.4.1 Discrete Time Planning

At the highest level of abstraction in our hierarchical control architecture, we let each individual vehicle be in a particular square in the Cartesian grid for one unit of time. For safety reasons, two vehicles in a square are forbidden since this case could potentially result in a "near miss" or a collision. We let the terminal, or goal point, have coordinates (0,0), and we give the grid a metric based on the norm $||(a, b)|| = |a| + |b|$, which in the air traffic control literature is known as the Manhattan metric.

When a vehicle enters the region around the terminal, it should trace a trajectory of minimal distance to the terminal, while avoiding contact with all other vehicles. Thus, if the vehicle is at (7,-5), it must reach the terminal in 12 time units. At the next instant of time, it must be at either (6,-5) or at (7,-4). We can denote this by the following controlled equation

$$(x_k, y_k) = (x_{k-1}, y_{k-1}) + (\delta_1, \delta_2),$$

where $|\delta_i| = 1$ or 0, and the following rules apply: $\delta_1 \delta_2 = 0$, $|x_k| \leq |x_{k-1}|$, $|y_k| \leq |y_{k-1}|$, and $\delta_1^2 + \delta_2^2 = 1$. These rules apply any time the position of the vehicle is described in terms of its location with respect to the terminal.

It is straightforward to see that the conditions concerning blockages surrounding the terminal can be completely characterized by the following three examples. The first case is displayed in Figure 10.4(a). It is the simplest blockage case, where it is impossible to move without x_2 being in conflict with either x_1 or x_3. The real problem is that all three vehicles are exactly three time units from the terminal, and there are only two possible routes for them to take.

The fact that there are exactly four directions from which the robots can enter the terminal means that at most four vehicles may enter at any given instant of time. Thus, if five or more vehicles are anywhere in the grid, equidistant from the terminal, there is an unavoidable blockage. This is illustrated in Figure 10.4(b). Similarly, if there are four vehicles equidistant from the terminal in two adjacent quadrants, this will force four vehicles to reach the terminal simultaneously from three different directions, as illustrated in Figure 10.4(c).

Can any other conflicts occur? The answer is no, and thus only the conflicts illustrated in Figures 10.4(a)–(c) lead to inevitable conflicts. This can be shown by working backward. Suppose that vehicles x and y are in conflict at time $k + 1$. Then at time k they were in a position where they had no choice on their next move. This requires that either one or the other was heading straight toward the terminal, or that there were other vehicles on the same level line. Thus either we are in the situation of Figure 10.4(a) as

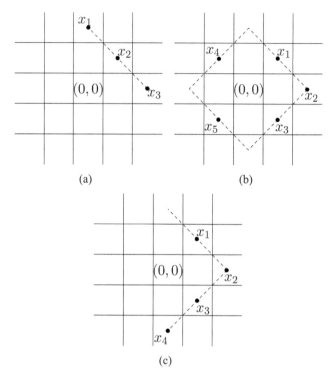

Figure 10.4 The three possible blockage cases.

a consequence of the situation in Figure 10.4(b), or we are in the situation of Figure 10.4(c), with several vehicles on the same level line in a single quadrant.

A safe route can thus be planned (and hence $\mathbf{q} \in P_{X_2}$ can be generated in the hierarchical control strategy) in a straightforward manner using standard graph geodesics if and only if the following conditions are satisfied:

1. *No more than two vehicles in a single quadrant are equidistant from the origin.*

2. *No more than three vehicles in two adjacent quadrants are equidistant from the origin.*

3. *No more then four vehicles are equidistant from the origin.*

10.4.2 Path Planning Using Monotone Splines

Assume that we are given a set of safe shortest-distance points through the grid, $\xi_i,\ i = 1, \ldots, N$, which constitute one instance of a conflict-free path in a multi-vehicle scenario. Furthermore, these points are located in the middle of squares in the grid, $\xi_i \in \mathcal{G}_i$, $i = 1, \ldots, N$, with side length 1. Thus $\mathcal{G}_i$ has midpoint (ξ_{xi}, ξ_{yi}), $i = 1, \ldots, N$. What we want to achieve is to generate a feasible route that *goes through only these specified squares in the grid.* We want to do this at the same time as we want to prevent the trajectories from oscillating, due to the fact that we ultimately want the vehicles to be able to follow the paths.

Without loss of generality, we assume that we are working with grid squares in the third quadrant, and the task is to drive the path through the squares to the terminal, located at the origin. This means that (ξ_{xi+1}, ξ_{yi+1}) is either $(\xi_{xi} + 1, \xi_{yi})$ or $(\xi_{xi}, \xi_{yi} + 1)$.

If we decouple the path planning problem into two subproblems, one along each axis, a preliminary version of the interpolation problem can be formulated as: Drive $(x(t), y(t)) \in \mathbb{R}^2$ close to (ξ_{xi}, ξ_{yi}) at the corresponding interpolation times, t_i, while staying in the specified grid squares for all times. At the same time we want the path to be smooth, which gives that we could, for instance, minimize the L_2-norm of the second derivatives in the x- and y-directions. In other words, given the system dynamics

$$\ddot{x}(t) = u_x(t), \; \ddot{y}(t) = u_y(t), \; u_x, u_y \in L_2[0, T],$$

what we want to do is minimize

$$\int_0^T \left(u_x(t)^2 + u_y(t)^2 \right) dt + \sum_{i=1}^N \left(\tau_{xi}(x(t_i) - \xi_{xi})^2 + \tau_{yi}(y(t_i) - \xi_{yi})^2 \right),$$

where T is the total time of the maneuver and $\tau_{xi} > 0$ and $\tau_{yi} > 0$ determine how much importance should be given to the waypoint fitting around (ξ_{xi}, ξ_{yi}). We furthermore impose the following constraint on our optimal control problem:

$$\forall t \in [0, T] \; \exists i \in \{1, \ldots, N\} \mid (x(t), y(t)) \in \mathcal{G}_i.$$

This last constraint is an infinite-dimensional constraint since it is a property that has to hold for all times, making the optimization problem very hard to solve [64]. Instead we would like to reformulate this as a problem where the grid constraints are finite. This can be achieved if we require that the curve is monotonously increasing in both the x- and y-directions (since we are in the third quadrant) while demanding that $(x(t_i), y(t_i)) \in \mathcal{G}_i, i = 1, \ldots, N$, which allows us to trade one infinite-dimensional constraint for another, that is, that $\dot{x}(t), \dot{y}(t) \geq 0 \; \forall t \in [0, T]$.

We can thus reformulate the problem as

$$\min_{u_x, u_y} \int_0^T (u_x^2(t) + u_y^2(t)) dt + \sum_{i=1}^N \left(\tau_{xi}(x(t_i) - \xi_{xi})^2 + \tau_{yi}(y(t_i) - \xi_{yi})^2 \right),$$

subject to

$$\ddot{x}(t) = u_x, \quad \ddot{y}(t) = u_y, \ u_x, u_y \in L_2[0,T],$$
$$\dot{x}(t) \geq 0, \quad \dot{y}(t) \geq 0, \quad \forall t \in [0,T],$$
$$(x(t_i), y(t_i)) \in \mathcal{G}_i, \ i = 1, \ldots, N,$$

where we have assumed that $x(0), y(0) \in \mathcal{G}_1, \dot{x}(0), \dot{y}(0) \geq 0$.

In [34] as well as in Chapter 7, it was found that the set of controls in $L_2[0,T] \times L_2[0,T]$ that satisfy the constraints is a closed, convex, and nonempty set, which is a strong enough result to guarantee the existence of a unique optimal solution. And, we now also know that the optimal control in $L_2[0,T] \times L_2[0,T]$ is piecewise linear. Furthermore, u_x only changes from different linear cases at the waypoints, or at times when $\dot{x}(t) = 0$, and similarly for u_y.

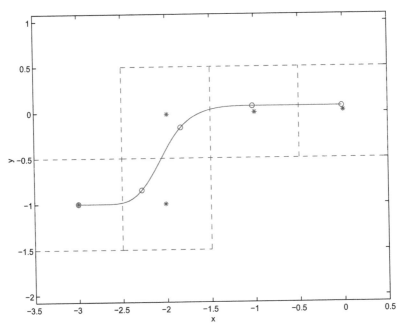

Figure 10.5 A path planning example corresponding to the solution in Figure 10.6. The circles and stars correspond to the actual positions at the interpolation times and the waypoints, respectively.

Based on these two facts it is possible to find the optimal piecewise linear control inputs in a computationally feasible way by solving a dynamic programming problem along each axis. An example of applying this method to the problem of planning planar, feasible paths are shown in Figures 10.5–10.6.

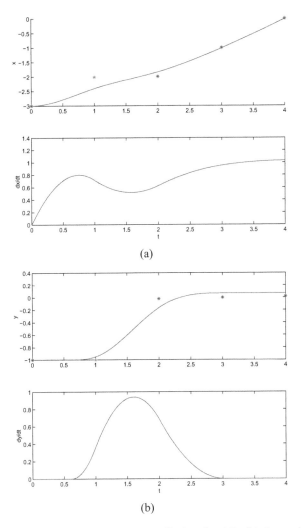

Figure 10.6 In (a), x (top) and $\dot{x}$ (bottom) are displayed, while (b) shows the solution in the y-direction. The stars correspond to the waypoints, and the derivatives are non-negative for all times.

SUMMARY

In this chapter, we discussed some applications of the various smoothing splines developed in this book. In particular, using the stereographic projection, it was shown how to produce smoothing splines on the sphere. An application of this was given in the context of telemetry through the problem of tracking marine animals.

We also used monotone smoothing splines for planning collision-free paths around airports and established a direct correspondence between Bezier curves and control theoretic, smoothing splines.

Chapter Eleven

NODE SELECTION

In this chapter, a solution is presented to the problem of selecting sample points in an optimal fashion. These points are used for interpolation and smoothing procedures, and, in particular, we derive necessary optimality conditions for the sample points. An example is presented concerning generalized smoothing splines that illustrate the generality as well as the numerical feasibility of the proposed approach.

11.1 BACKGROUND

So far in this book, we have seen how to generate a large collection of different curves based on given sets of data points. If these points are to be selected rather than just given somehow, one can of course ask how they should be selected. This turns out to be a question that cannot be tackled as a minimum norm problem in a Hilbert space. Nonetheless, it is such an important problem that we include it in this book for the sake of completeness.

In particular, we consider the problem of selecting the data points in an optimal fashion for interpolating or smoothing procedures. What the solution to this problem entails is the computation of how trajectories from switched, autonomous dynamical systems depend on the sample times. For this, optimal timing control, based on classic variational techniques, will be employed in order to find locally optimal sample times.

Historically, the main explicit result relating the selection of interpolation (or sample) times to the performance of the resulting curve, is given by the *Tschebyscheff polynomials*.

Given a function $h(t) \in C^{N-1}(t_0, t_f)$, the unique polynomial P_{N-1} that interpolates the data points $h(t_1), \ldots, h(t_N)$ satisfies

$$|h(t) - P_{N-1}(t)| \le \max_{t_0 \le \xi \le t_f} |h^N(\xi)| \max_{t_0 \le \xi \le t_f} \frac{\prod_{i=1}^{N} |\xi - t_i|}{N!}$$

$$= H(t_1, \ldots, t_N)$$

as shown in [22].

Moreover, the solution to the problem

$$\min_{t_1,\ldots,t_N} H(t_1,\ldots,t_N)$$

is given by the Tschebyscheff polynomials. However, this result only holds for exact polynomial interpolation. Since we are interested in solving a more general problem, with general curve and cost types, the Tschebyscheff polynomials will unfortunately not provide much assistance in this quest.

11.2 SAMPLING FOR INTERPOLATION AND SMOOTHING

The connection between data interpolation (and smoothing) and optimal control has been made repeatedly in this book. (See also, for example, [66],[85],[96].)

However, regardless what method is used for generating the curve, the optimal control vantage point lets us assume that, from a general point of view, the underlying system dynamics is given by $\dot{x} = F(x, u)$, where $x \in \mathbb{R}^n, u \in \mathbb{R}^q$, and F is smooth. Independent of interpolating or smoothing procedure, the resulting control law will in general depend on time t as well as the sample times $\tau = (\tau_1,\ldots,\tau_N)^T \in \mathbb{R}^N$. Moreover, u will not be smooth (or even continuous) at the sample times, while it will be smooth for all other times. We can thus let the resulting optimal control law be given by

$$u(t,\tau) = G_i(t,\tau), \quad \forall t \in [\tau_{i-1},\tau_i),$$

where $i = 1,\ldots,N+1$, $\tau_0 = t_0$, and $\tau_{N+1} = t_f$. In other words, the now autonomous yet switched system is given by

$$\dot{x} = F(x, G_i(t,\tau)) = f_i(x,t,\tau), \quad \forall t \in [\tau_{i-1},\tau_i).$$

Moreover, if we assume that the data points are generated from an underlying curve $h(t) \in \mathbb{R}$, and if we let the output from the dynamical system be $y(t) = g(x(t)) \in \mathbb{R}$, we can define $L(x(t), t)$ as

$$L(x(t),t) = (g(x(t)) - h(t))^2,$$

and try to minimize the cost:

$$\min_{\tau} J(\tau) = \min_{\tau} \int_{t_0}^{t_f} L(x(t),t)dt,$$

subject to

$$\dot{x} = f_i(x,t,\tau), \ t \in [\tau_{i-1},\tau_i), \ i = 1,\ldots,N+1$$
$$x(t_0) = x_0.$$

This general timing control problem will be addressed in the next section, followed by a discussion about how the results should be used when producing generalized smoothing splines. It should, however, be noted that if the system dynamics $f_i(x, t, \tau)$ did not depend on the switching times τ explicitly, this would be a standard timing control problem that has been solved, for example, in [35],[86],[92],[98].

In fact, in [35], it was found that the gradient of the cost (if the dynamics does not depend on the switching times) was given by

$$\frac{dJ}{d\tau_i} = \lambda(\tau_i)\Big(f_i(x(\tau_i)) - f_{i+1}(x(\tau_i))\Big),$$

given the costate equation

$$\begin{aligned}\dot{\lambda} &= -\frac{\partial L}{\partial x} - \lambda\frac{\partial f_i}{\partial x}, \ t \in [\tau_{i-1}, \tau_i),\\ \lambda(t_f) &= 0.\end{aligned}$$

Hence, the task undertaken in this chapter is in part to extend this result to the case when f_i does depend on τ.

11.3 OPTIMAL TIMING CONTROL

As before, consider the autonomous, switched dynamical system

$$\begin{aligned}\dot{x} &= f_i(x, t, \tau), \ t \in [\tau_{i-1}, \tau_i),\\ x(t_0) &= x_0,\end{aligned} \tag{11.1}$$

where $\{f_i(x, t, \tau)\}_{i=1}^{N+1}$ is a given sequence of smooth mappings from $\mathbb{R}^n \times \mathbb{R} \times \mathbb{R}^N$ to $\mathbb{R}^n$. Moreover, let L be a smooth function from $\mathbb{R}^n \times \mathbb{R} \to \mathbb{R}$, and let the cost J be given as before by

$$J(\tau) = \int_{t_0}^{t_f} L(x(t), t)dt.$$

Note that J may very well be nonconvex, which means that only local optima can be expected to be obtained from gradient-based algorithms. The computation of the gradient of J with respect to the switching times is the contribution in this section; it will be based on the classic variational approach where the dynamical constraints are adjoined to the cost function via the costate variable λ.

11.3.1 Gradient Computation

We have

$$
J_0 = \sum_{i=1}^{N+1} \left(\int_{\tau_{i-1}}^{\tau_i} \left(L(x(t),t) + \lambda(f_i(x(t),t,\tau) - \dot{x}) \right) dt \right).
$$

By $i(t)$ we understand that $i(t) = j$ when $t \in [\tau_{j-1}, \tau_j)$, and we consider the variation in J due to a perturbation in τ_k only. Hence, we replace $f_i(x,t,\tau)$ with $f_i(x,t,\tau_k)$ for the sake of notational ease. The variation is obtained through the small perturbation $\tau_k \to \tau_k + \epsilon\theta_k$, where $\epsilon \ll 1$, which results in the state variation $x \to x + \epsilon\eta$. The cost function for the perturbed system is

$$
\begin{aligned}
J_\epsilon &= \int_{t_0}^{\tau_k} \Big[L(x+\epsilon\eta, t) + \lambda(f_{i(t)}(x+\epsilon\eta, t, \tau_k + \epsilon\theta_k) - \dot{x} - \epsilon\dot{\eta}) \Big] dt \\
&\quad + \int_{\tau_k}^{\tau_k+\epsilon\theta_k} \big[L(x+\epsilon\eta, t) + \lambda(f_k(x+\epsilon\eta, t, \tau_k + \epsilon\theta_k) - \dot{x} - \epsilon\dot{\eta}) \big] dt \\
&\quad + \int_{\tau_k+\epsilon\theta_k}^{t_f} \Big[L(x+\epsilon\eta, t) + \lambda(f_{i(t)}(x+\epsilon\eta, t, \tau_k + \epsilon\theta_k) - \dot{x} - \epsilon\dot{\eta}) \Big] dt.
\end{aligned}
$$

A first order approximation of the continuously differentiable functions f_i and L gives

$$
\begin{aligned}
J_\epsilon - J_0 &= \int_{t_0}^{t_f} \left(\frac{\partial L}{\partial x}\epsilon\eta + \lambda\left(\frac{\partial f_{i(t)}}{\partial x}\epsilon\eta + \frac{\partial f_{i(t)}}{\partial \tau_k}\epsilon\theta_k - \epsilon\dot{\eta} \right) \right) dt \\
&\quad + \int_{\tau_k}^{\tau_k+\epsilon\theta_k} \lambda(f_k(x,t,\tau_k) - f_{k+1}(x,t,\tau_k)) dt.
\end{aligned}
$$

Following the development in [35], we choose the continuous costate

$$
\begin{aligned}
\dot{\lambda} &= -\frac{\partial L}{\partial x} - \lambda\frac{\partial f_i}{\partial x}, \quad t \in [\tau_{i-1}, \tau_i), \\
\lambda(t_f) &= 0,
\end{aligned} \tag{11.2}
$$

which, through integration by parts, simplifies the variation $\delta J = (J_\epsilon - J_0)/\epsilon$ to

$$
\delta J = \left(\int_{t_0}^{t_f} \lambda\frac{\partial f_{i(t)}}{\partial \tau_k} dt + \lambda(\tau_k)(f_k - f_{k+1})\Big|_{t=\tau_k} \right) \theta_k.
$$

We thus have that the kth component of the gradient of J with respect to τ is given by

$$
\frac{dJ}{d\tau_k} = \int_{t_0}^{t_f} \lambda\frac{\partial f_{i(t)}}{\partial \tau_k} dt + \lambda(\tau_k)(f_k - f_{k+1})\Big|_{t=\tau_k}, \tag{11.3}
$$

which allows us to use gradient-based algorithms for selecting locally optimal sample points in our interpolation and smoothing problems, as we will see in the next section. We start by a brief discussion of the numerical aspects of this approach.

11.3.2 Gradient Descent

The reason why the formula derived in the previous paragraphs is particularly easy to work with is that it gives us access to a very straightforward numerical algorithm.

For each iteration k, let $\tau(k)$ be the set of switching times, and compute the following:

1. Compute $x(t)$ forward in time on $[t_0, t_f]$ by integrating (11.1) from $x(t_0) = x_0$.

2. Compute $\lambda(t)$ backward in time from t_f to t_0 by integrating (11.2) from $\lambda(t_f) = 0$.

3. Use (11.3) to compute $dJ/d\tau(\tau(k))$.

4. Update τ as

$$\tau(k+1) = \tau(k) - l(k) \left(\frac{dJ}{d\tau}(\tau(k)) \right)^T,$$

   where $l(k)$ is the stepsize, e.g., given by the Armijo algorithm [8].

5. Repeat.

Note that this method will only converge to a local minimum. But, as we will see, it can still give quite significant reductions in cost.

11.3.3 Example - Linear Approximations

In this example, we try to approximate a continuous function $h : [t_0, t_f] \to \mathbb{R}$ by a function x such that, for $i = 1, \ldots, N+1$, $\forall t \in [\tau_{i-1}, \tau_i)$,

$$x(t) = h(\tau_{i-1}) + (t - \tau_{i-1}) \frac{h(\tau_i) - h(\tau_{i-1})}{\tau_i - \tau_{i-1}},$$

where $\tau_0 = t_0$ and $\tau_{N+1} = t_f$. This autonomous switched system is simpler than the general case considered previously since the derivative function

$\dot{x}(t) = f_i(x(t), \tau)$ on $[\tau_{i-1}, \tau_i)$ here only depends on τ_{i-1} and τ_i, that is,

$$\dot{x}(t) = f_i(\tau_{i-1}, \tau_i) = \frac{h(\tau_i) - h(\tau_{i-1})}{\tau_i - \tau_{i-1}} \quad \text{on } [\tau_{i-1}, \tau_i).$$

We now apply the algorithm to the problem of determining $\tau_1, \ldots, \tau_N$ in order to minimize the cost function

$$J(\tau)) = \int_{\tau_0}^{\tau_N+1} (h(t) - x(t))^2 dt.$$

Figure 11.1 shows how the algorithm converges. The following parameters were used:

$$\begin{cases} h(t) = 5\sin\left(\frac{2\pi t}{300}\right) + 3\sin\left(\frac{2\pi t}{100}\right) + \frac{t^2}{20000} - \frac{t}{50}, \\ [t_0, t_f] = [0, 200], \quad N = 4, \quad l = 1. \end{cases}$$

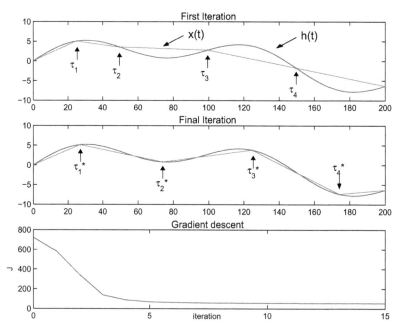

Figure 11.1 Optimal linear approximation.

The lowest figure in Figure 11.1 shows how fast the algorithm converges. The optimal solution is reached after a very few iterations, in spite of a "bad" initial guess and a constant step size l.

11.4 APPLICATIONS TO SMOOTHING SPLINES

The optimal selection of nodes is common in such problems as the air traffic control problem. If an airplane receives the command to be at 10,000 ft in $t+$ 2 minutes, and if this would violate acceleration constraints, the command would be changed to, for example, $t+4$ minutes–an optimal selection of the node.

As discussed previously, we can view the smoothing problem as a problem of finding the optimal control that drives the output of a given linear control system close to given data points. In particular, given the dynamics

$$\dot{x} = Ax + bu, \ x \in \mathbb{R}^n,$$
$$y = cx,$$
$$(11.4)$$

the (unique) optimal solution to the problem

$$\min_{u \in L_2[t_0, t_f]} \int_{t_0}^{t_f} u^2(t)dt + \sum_{i=1}^{N} w_i(y(\tau_i) - \xi_i)^2, \qquad (11.5)$$

was in Chapter 4 found to be

$$u(t) = \ell(t)^T (I + \mathcal{W}G)^{-1}\mathcal{W}\xi,$$

where

$$\mathcal{W} = \begin{pmatrix} w_1 & 0 & \cdots & 0 \\ 0 & w_2 & \cdots & 0 \\ \vdots & \vdots & \ddots & \vdots \\ 0 & 0 & \cdots & w_N \end{pmatrix}, \ \xi = \begin{pmatrix} \xi_1 \\ \xi_2 \\ \vdots \\ \xi_N \end{pmatrix} = \begin{pmatrix} h(\tau_1) \\ h(\tau_2) \\ \vdots \\ h(\tau_N) \end{pmatrix},$$

$$\ell(t) = \begin{pmatrix} \ell_1(t) \\ \ell_2(t) \\ \vdots \\ \ell_N(t) \end{pmatrix}, \ \ell_i(t) = \begin{cases} ce^{A(\tau_i - t)}b & \text{if } t \leq \tau_i, \\ 0 & \text{otherwise,} \end{cases}$$

and where the Grammian G is given by

$$G = \int_{t_0}^{t_f} \ell(s)\ell(s)^T ds \in \mathbb{R}^{N \times N}.$$

Note that the definition of the basis functions $\ell_i(t)$ implies that u may be discontinuous at τ_i. In fact, we could define a new set of basis functions

$$
\zeta_i(t) = \begin{pmatrix} 0 \\ \vdots \\ 0 \\ ce^{A(\tau_i - t)}b \\ ce^{A(\tau_{i+1} - t)}b \\ \vdots \\ ce^{A(\tau_N - t)}b \end{pmatrix} , \ t \in [\tau_{i-1}, \tau_i), \ i = 1, \ldots, N,
$$

with $\zeta_{N+1} = 0$. Hence we have the new system

$$
\begin{aligned}
\dot{x} &= Ax + bu \\
&= Ax + b\zeta_i^T(t, \tau)(I + \mathcal{W}G(\tau))^{-1}\mathcal{W}\xi(\tau) \\
&= f_i(x, t, \tau), t \in [\tau_{i-1}, \tau_i), \\
y &= cx = g(x),
\end{aligned}
$$

that is in the prescribed form.

In order to be able to apply the gradient-based optimization methods, we need to obtain expressions for $\partial L/\partial x$, $\partial f_i/\partial x$, and $\partial f_i/\partial \tau_k$. If, as before, we let L be given by $(y(t) - h(t))^2$, we get for $i = 1, \ldots, N+1$

$$
\frac{\partial f_i}{\partial x} = A,
$$

$$
\frac{\partial L}{\partial x} = 2c(cx(t) - h(t)),
$$

$$
\frac{\partial f_i}{\partial \tau_k} = b\ell(t)^T (I + \mathcal{W}G)^{-1} \mathcal{W}\delta\xi_k + b\delta\ell_k(t)^T \mathcal{W}(I + \mathcal{W}G)^{-1}\xi
$$

$$
- b\ell(t)^T (I + \mathcal{W}G)^{-1} \mathcal{W} \frac{\partial G}{\partial \tau_k} (I + \mathcal{W}G)^{-1}\mathcal{W}\xi,
$$

where

$$
\delta\xi_k = \begin{pmatrix} 0 \\ \vdots \\ 0 \\ \frac{\partial h}{\partial t}(\tau_k) \\ 0 \\ \vdots \\ 0 \end{pmatrix} \quad \leftarrow \ k\text{th position}
$$

$$\delta\ell_k(t) = \begin{pmatrix} 0 \\ \vdots \\ 0 \\ \frac{\partial \ell_k}{\partial \tau_k}(t) \\ 0 \\ \vdots \\ 0 \end{pmatrix} \quad \leftarrow \; k\text{th position}$$

$$\frac{\partial \ell_k}{\partial \tau_k}(t) = \begin{cases} cAe^{A(\tau_k - t)}b & \text{if } t \leq \tau_k, \\ 0 & \text{otherwise}, \end{cases}$$

$$\frac{\partial G}{\partial \tau_k} = \int_{t_0}^{t_f} \left(\ell(s)\delta\ell_k(s)^T + \delta\ell_k(s)\ell(s)^T \right) ds.$$

Note that for this system $(f_k - f_{k+1})|_{t=\tau_k} = 0$, which simplifies the derivative of the cost to

$$\frac{dJ}{d\tau_k} = \int_{t_0}^{t_f} \lambda \frac{\partial f_{i(t)}}{\partial \tau_k} dt.$$

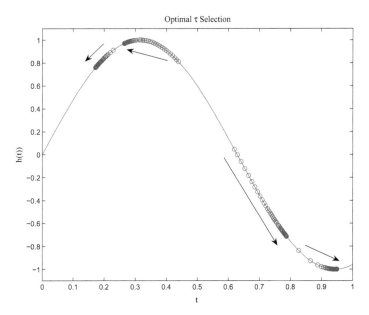

Figure 11.2 Movements of the sample times when creating smoothing splines for the underlying curve $h(t) = \sin(5t)$.

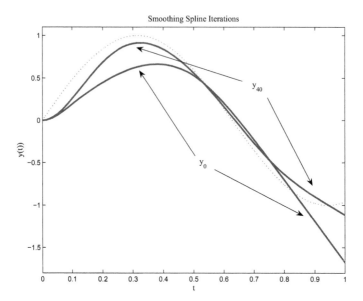

Figure 11.3 Smoothing splines (solid) obtained at the first and the 40th iterations, together with the underlying curve (dotted).

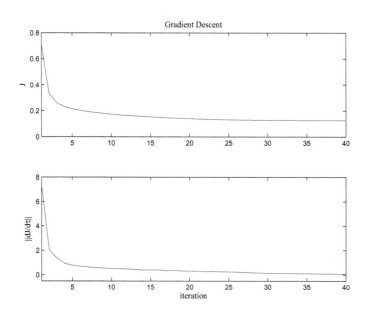

Figure 11.4 Evolution of $J(\tau(k))$ together with $\|dJ/d\tau(\tau(k))\|$ as a function of k, $k = 1, \ldots, 40$.

11.4.1 Example

In this paragraph we apply the node selection method to the system

$$A = \begin{pmatrix} 0 & 1 & 0 \\ 0 & 0 & 1 \\ 0 & 0 & 0 \end{pmatrix}, \ B = \begin{pmatrix} 0 \\ 0 \\ 1 \end{pmatrix}, \ C = (1,0,0),$$

which gives the standard quintic smoothing spline.

Results from applying the gradient descent method using the Armijo step-size over 40 iterations is shown in Figures 11.2–11.4. In that example, the underlying curve was given by $h(t) = \sin(5t)$, and four sample times where selected with $w_i = 1$, $i = 1, \ldots, 4$.

SUMMARY

In this chapter we presented a method where variational techniques were employed to select optimal sample points for interpolation and smoothing applications. This method, moreover, resulted in a numerically straightforward algorithm that was put to use in the context of generalized smoothing splines. It produced results that go well beyond the previously known results on Tschebyscheff polynomials.

Bibliography

[1] J. Ackermann. *Robust Control*. Springer-Verlag, London, 1993.

[2] M.B. Adams, A.S. Willsky, and B.C. Levy. Linear estimation of boundary value stochastic processes. I. The role and construction of complementary models. *IEEE Transactions on Automatic Control*, 29(9):803–811, 1984.

[3] M.B. Adams, A.S. Willsky, and B.C. Levy. Linear estimation of boundary value stochastic processes. II. 1-D smoothing problems. *IEEE Transactions on Automatic Control*, 29(9):811–821, 1984.

[4] N. Agwu. *Optimal control of dynamic systems and its application to spline approximation*. Dissertation, Texas Tech University, 1996.

[5] N. Agwu and C. Martin. Optimal control of dynamic systems: Application to spline approximations. *Applied Mathematics and Computation*, 97:99–138, 1998.

[6] A. Ailon and R. Segev. Driving a linear constant system by piecewise constant control. *International Journal of Control*, 47(3):815–825, 1988.

[7] C. Altafini. The De Casteljau algorithm on SE(3). In A. Isidori, F. Lamnabhi-Lagarrigue, and W. Respondek, editors, *Nonlinear Control in the Year 2000*. Springer, New York, 2000.

[8] L. Armijo. Minimization of functions having Lipschitz continuous first-partial derivatives. *Pacific Journal of Mathematics*, 16:1–3, 1966.

[9] A.R. Barron and C.H. Sheu. Approximation of density functions by sequences of exponential families. *Annals of Statistics*, 19, 1991.

[10] A.E. Bryson and Y.C. Ho. *Applied Optimal Control*. Wiley, New York, 1975.

[11] G. Bunnett, P. Crouch, and F. Silva Leite. Spline elements on spheres. In M. Daehlen, T. Lynch, and L. Schumaker, editors, *Mathematical Methods for Curves and Surfaces*, pages 49–54. Vanderbilt University Press, Nashville, TN, 1995.

[12] G. Campion, G. Bastin, and B. D'Andréa-Novel. Structural properties and classification of kinematic and dynamic models of wheeled mobile robots. *IEEE Transactions on Robotics and Automation*, 12(1), Feb. 1996.

[13] R.J. Carroll, P. Hall, T.V. Apanasovich, and X. Lin. Histospline method in nonparametric regression models with application to clustered/longitudinal data. *Statistica Sinica*, 14(3):649–674, 2004.

[14] CCC/STSL education program–sea turtle migration-tracking introduction. http://cccturtle.org/satintro.htm.

[15] D. Chauveau, C.F. Martin, A.C.M. van Rooij, and F.H. Ruymgaart. Discrete signed mixtures of exponentials. *Communications in Statistics. Stochastic Models*, 12(2):245–263, 1996.

[16] R. Chesser. personal communication.

[17] P. Constantini. Boundary-valued shape-preserving interpolating splines. *ACM Transactions on Mathematical Software*, 23(2):229–251, 1997.

[18] P. Crouch, G. Kun, and F. Silva Leite. Generalization of spline curves on the sphere: A numerical comparison. In *Proceedings of the Third Portuguese Conference on Automatic Control*, volume 2, pages 474–451, Coimbra, Portugal, September 1998.

[19] P. Crouch, G. Kun, and F. Silva Leite. The De Casteljau algorithm on Lie groups and spheres. *Journal of Dynamical and Control Systems*, 5(3):397–429, 1999.

[20] P. Crouch, F. Silva Leite, and G. Kun. Geometric splines. In *14th IFAC World Congress*, pages 533–538, Beijing, China, July 1999.

[21] R.B. Darst and S. Sahab. Approximation of continuous and quasi-continuous functions by monotone functions. *Journal of Approximation Theory*, 38:9–27, 1983.

[22] P.J. Davis. *Interpolation and Approximation*. Dover, New York, 1975.

[23] P.J. Davis and P. Rabinowitz. *Methods of Numerical Integration*. Academic Press, New York, 1975.

[24] C. de Boor. *A Practical Guide to Splines*. Springer, Berlin, 1978.

[25] C. Canudas de Wit. Trends in mobile robot and vehicle control. In B. Siciliano and K.P. Valavanis, editors, *Control Problems in Robotics*, pages 151–176. Springer, London, 1998.

[26] C. Canudas de Wit, B. Siciliano, and G. Bastin. *Theory of Robot Control*. Springer, New York, 1996.

[27] J. Desai, J. Ostrowski, and V. Kumar. Controlling formations of multiple mobile robots. In *IEEE International Conference on Robotics and Automation*, pages 2864–2869, Leuven, Belgium, May 1998.

[28] B. Doolin and C. Martin. *Differential Geometry for Control Engineers*. Marcel Dekker, New York, 1990.

[29] M. Egerstedt, X. Hu, and A. Stotsky. Control of a car-like robot using a virtual vehicle approach. In *IEEE Conference on Decision and Control*, pages 1502–1507, Tampa, Florida, December 1998.

[30] M. Egerstedt and C. Martin. Trajectory planning for linear control systems with generalized splines. In *Mathematical Theory of Networks and Systems*, pages 999–1002, Padova, Italy, 1998.

[31] M. Egerstedt and C. Martin. Optimal control and monotone smoothing splines. In *New Trends in Nonlinear Dynamics and Control, and Their Applications*, pages 279–294, Berlin, 2003. Springer.

[32] M. Egerstedt and C. Martin. Statistical estimates for generalized splines. *ESAIM*, 9:553–562, August 2003.

[33] M. Egerstedt, C. Martin, and Y. Zhou. Optimal approximation of functions. *Communication in Information and Systems*, 1(1):101–112, 2001.

[34] M. Egerstedt and C.F. Martin. Monotone smoothing splines. In *Mathematical Theory of Networks and Systems*, pages 1020–1028, Perpignan, France, 2000.

[35] M. Egerstedt, Y. Wardi, and H. Axelsson. Transition-time optimization for switched systems. *IEEE Transactions on Automatic Control*, 51(1):110–115, January 2006.

[36] R.L. Eubank. *Nonparametric regression and spline smoothing*, volume 157 of *Statistics: Textbooks and Monographs*. Marcel Dekker, New York, 1999.

[37] G. Farin. *Curves and Surfaces for Computer-Aided Geometric Design: A Practical Guide, 4th Ed.* Academic Press, San Diego, CA, 1997.

[38] G. Farin. *NURBS: From Projective Geometry to Practical Use, 2nd Edition*. A.K. Peters, Inc., Natick, MA, 1999.

[39] J. Fax and R. Murray. Information flow and cooperative control of vehicle formations. *IEEE Transactions on Automatic Control*, 49:1465–1476, Sept. 2004.

[40] G. Ferrari-Trecate, M. Egerstedt, A. Buffa, and M. Ji. Laplacian sheep: a hybrid, stop-go policy for leader-based containment control. In *Hybrid Systems: Computation and Control*, pages 212–226, Santa Barbara, CA, 2006.

[41] W.T. Ford. On interpolation and approximation by polynomials with monotone derivatives. *Journal of Approximation Theory*, 10:123–130, 1974.

[42] W. Gautschi. On inverses of vandermonde and confluent vandermonde matrices. *Numerische Mathematik*, 4:117–123, 1962.

[43] B. K. Ghosh and E. P. Loucks. A perspective theory for motion and shape estimation in machine vision. *SIAM Journal of Control and Optimization*, 33(5):1530–1559, September 1995.

[44] C. Godsil and G. Royle. *Algebraic Graph Theory*. Springer.

[45] P. Hall and J.D. Opsomer. Theory for penalised spline regression. *Biometrika*, 92(1):105–118, 2005.

[46] M.H. Hansen and C. Kooperberg. Spline adaptation in extended linear models. with comments and a rejoinder by the authors. *Statistcal Science*, 17(1):2–51, 2002.

[47] U. Hornung. Interpolation by smooth functions under restrictions on the derivatives. *Journal of Approximation Theory*, 28:227–237, 1980.

[48] K. Hüper and F. Silva Leite. Smooth interpolating curves with applications to path planning. In *10th Mediterranean Conference on Control and Automation*, pages 436–442, Lisbon, Portugal, July 2002.

[49] G.L. Iliev. Exact estimates for monotone interpolation. *Journal of Approximation Theory*, 28:101–112, 1980.

[50] A. Jadbabaie, J. Lin, and A. S. Morse. Coordination of groups of mobile autonomous agents using nearest neighbor rules. *IEEE Transactions on Automatic Control*, 48(6):988–1001, 2003.

[51] M. Ji and M. Egerstedt. A graph-theoretic characterization of controllability for multi-agent systems. In *American Control Conference*, 2007.

[52] M. Ji, A. Muhammad, and M. Egerstedt. Leader-based multi-agent coordination: Controllability and optimal control. In *American Control Conference*, pages 1358–1363, Minneapolis, MN, June 2006.

[53] H. Kano, M. Egerstedt, H. Nakata, and C.F. Martin. B-splines and control theory. *Applied Mathematics and Computation*, 145(2-3):265–288, 2003.

[54] H. Kano, H. Nakata, and C. Martin. Optimal curve fitting and smoothing using normalized uniform B-splines: a tool for studying complex systems. *Applied Mathematics and Computation*, 169(1):96–128, October 2005.

[55] M. Karasalo, X. Hu, and C.F. Martin. Localization and mapping using iterated smoothing splines. In *Proceedings of ICRA*, 2006.

[56] C. Kooperberg and C.J. Stone. A study of logspline density estimation. *Computational Statistics and Data Analysis*, 12(3):327–347, 1991.

[57] A.J. Krener. Boundary value linear systems. In *Systems Analysis Conference*, pages 149–165, Bordeaux, France, 1978.

[58] X. Li and J. Ramsay. Curve registration. *Journal of the Royal Statistical Society Series B Statistical Methodology*, 60(2):351–363, 1998.

[59] Z. Li and C. Martin. An inverse problem for a linear dynamical ssytem with noise. *Interantional Journal of Control*, 62:1291–1317, 1995.

[60] J. Lin, A. Morse, and B.D.O. Anderson. The multi-agent rendezvous problem. In *IEEE Conference on Decision and Control*, pages 1508–1513, Maui, Hawaii, Dec. 2003.

[61] Z. Lin, M. Broucke, and B. Francis.   Local control strategies for groups of mobile autonomous agents. *IEEE Transactions on Automatic Control*, 49(4):622–629, 2004.

[62] A. Lindquist and J. Sand. *An introduction to mathematical systems theory*. KTH Lecture Notes, Division of Optimization and Systems Theory, Royal Institute of Technology (KTH), 1996.

[63] Loggerhead Turtle Data/WhaleNet. Information is available online at http://whale.wheelock.edu/whalenet-stuff/StopGraysReef02.

[64] D.G. Luenberger. *Optimization by Vector Space Methods*. John Wiley & Sons, New York, 1969.

[65] E. Mammen and S. van de Geer. Locally adaptive regression splines. *Annals of Statistics*, 25(1):387–413, 1997.

[66] O.L. Mangasarin and L.L. Schumaker. Splines via optimal control. In I.J. Schoenberg, editor, *Approximation with Special Emphasis on Spline Functions*. Academic Press, New York, NY, 1969.

[67] C. Martin. Finite escape time for riccati differential equations. *Systems and Control Letters*, 1:121–131, 1981.

[68] C. Martin, P. Enqvist, J. Tomlinson, and Z. Zhang. Linear control theory, splines and interpolation. In J. Lund and K. Bowers, editors, *Computation and Control IV*, pages 269–288. Birkhauser, Berlin, 1995.

[69] C.F. Martin, S. Sun, and M. Egerstedt. Optimal control, statistics and path planning. *Mathematics and Computer Modelling*, 33(1-3):237–253, 2001.

[70] R.F. Martin. *Consumption, Durable Goods, and Transaction Costs*. Dissertation, University of Chicago, 2002.

[71] M. Mesbahi. State-dependent graphs. In *IEEE Conference on Decision and Control*, pages 3058–3063, Maui, Hawaii, Dec. 2003.

[72] J.J. Millspaugh and J. Marzluff (editors). *Radio Tracking and Animal Populations*. Academic Press, San Diego, CA, 2001.

[73] R. Murray and S. Sastry. Nonholonomic motion planning: Steering using sinusoids. *IEEE Transactions on Automatic Control*, 38(5):700–716, 1993.

[74] L. Noakes, G. Heinzinger, and B. Paden. Cubic splines on curved spaces. *IMA Journal of Mathematics, Control and Information*, 6:465–473, 1989.

[75] R. Olfati-Saber. Flocking for multi-agent dynamic systems: Algorithms and theory. *IEEE Transactions on Automatic Control*, 51(3):401–420, March 2006.

[76] E. Passow, L. Raymon, and J.A. Rouler. Comotone polynomial approximation. *Journal of Approximation Theory*, 11:221–224, 1974.

[77] E. Polak. *Algorithms and Consistent Approximations*. Springer, New York, 1997.

[78] I.G. Priede and S.M. Swift. *Wildlife Telemetry: Remote Monitoring and Tracking of Animals*. Ellis Horwood Series in Environmental Management, Science and Technology. Ellis Horwood, New York, 1993.

[79] A. Rahmani and M. Mesbahi. On the controlled agreement problem. In *American Control Conference*, pages 1376–1381, June 2006.

[80] J. Ramsay and B. Silverman. *Functional Data Analysis*. Springer Series in Statistics, Springer, New York, 1997.

[81] W. Ren and R. Beard. Consensus of information under dynamically changing interaction topologies. In *American Control Conference*, pages 4939–4944, Boston, MA, June 2004.

[82] R.T Rockafellar. *Convex Analysis*. Princeton University Press, Princeton, NJ, 1970.

[83] W. Rudin. *Principles of Mathematical Analysis*. McGraw-Hill, New York, 1976.

[84] N. Sarkar, X. Yun, and V. Kumar. Dynamic path following: A new control algorithm for mobile robots. In *32nd Conference on Decision and Control*, pages 722–728, San Antonio, Texas, Dec. 1993.

[85] L.L. Schumaker. *Spline Functions: Basic Theory*. John Wiley and Sons, New York, 1981.

[86] M.S. Shaikh and P. Caines. On trajectory optimization for hybrid systems: Theory and algorithms for fixed schedules. In *IEEE Conference on Decision and Control*, Las Vegas, NV, Dec. 2002.

[87] B.W. Silverman. Spline smoothing: the equivalent variable kernel method. *Annals of Statistics*, 12:898–916, 1984.

[88] J. Smith and C. Martin. Approximation, interpolation and sampling. In *Differential Geometry: The Interface Between Pure and Applied Mathematics*, pages 227–252, San Antonio, Texas, 1986.

[89] L.M. Smith. *Playas of the Great Plains*. University of Texas Press, Austin, 2003.

[90] S. Sugathadasa, C. Martin, and W.P. Dayawansa. Convergence of the extended Kalman filter to locate a moving target in wild life telemetry. In *IEEE Conference on Decision and Control*, Sydney, Australia, Dec. 2000.

[91] S. Sun, M. Egerstedt, and C.F. Martin. Control theoretic smoothing splines. *IEEE Transactions on Automatic Control*, 45(12):2271–2279, 2000.

[92] H.J. Sussmann. Set-valued differentials and the hybrid maximum principle. In *IEEE Conference on Decision and Control*, pages 558–563, Sydney, Australia, 2000.

[93] H. Tanner, A. Jadbabaie, and G. Pappas. Stable flocking of mobile agents, part II : Dynamic topology. In *IEEE Conference on Decision and Control*, pages 2016–2021, Maui, Hawaii, Dec. 2003.

[94] H.G. Tanner. On the controllability of nearest neighbor interconnections. In *IEEE Conference on Decision and Control*, pages 2467–2472, Atlantis, Bahamas, 2004.

[95] H.G. Tanner, G.J. Pappas, and V. Kumar. Leader-to-formation stability. *IEEE Transactions on Automatic Control*, 20(3):433–455, 2004.

[96] G. Wahba. Spline models for observational data. In *CBMS-NSF Regional Conference Series in Applied Mathematics, 59. Society for Industrial and Applied Mathematics (SIAM)*, Philadelphia, PA, 1990.

[97] E. Wegman and I. Wright. Splines in statistics. *Journal of the American Statistical Association*, 78:351–365, 1983.

[98] X. Xu and P. Antsaklis. Optimal control of switched autonomous systems. In *IEEE Conference on Decision and Control*, pages 1422–1428, Las Vegas, NV, Dec. 2002.

[99] M. Zefran, V. Kumar, and C. Corke.   On the generation of smooth three-dimensional rigid body motions. *IEEE Transactions on Robotics and Automation*, 12(4):576–589, 1998.

[100] Z. Zhang, J. Tomlinson, and C. Martin.   Splines and linear control theory. *Acta Applicandae Mathematicae*, 49:1–34, 1997.

[101] Y. Zhou, W. Dayawansa, and C. Martin. Control theoretic smoothing splines are approximate linear filters. *Communications in Information and Systems*, 4:253–272, 2004.

[102] Y. Zhou, M. Egerstedt, and Martin.   Control theoretic splines with deterministic and random data. In *IEEE Conference on Decision and Control*, Seville, Spain, Dec. 2005.

[103] Y. Zhou, M. Egerstedt, and C. Martin.   Optimal trajectories between affine subspaces. Submitted.

[104] Y. Zhou, M. Egerstedt, and C. Martin.   Optimal approximation of functions. *Communications in Information and Systems*, 1(1):99–110, January 2001.

[105] Y. Zhou, M. Egerstedt, and C. Martin.   Hilbert space methods for control theoretic splines: A unified treatment. *Communications in Information and Systems*, 6(1):55–82, 2006.

[106] Y. Zhou and C. Martin.   A regularized solution to the Birkhoff interpolation problem. *Communication in Information and Systems*, 4:89–102, 2004.

Index

Milton Keynes UK
Ingram Content Group UK Ltd.
UKHW020134180824
447066UK00003B/52